CHRISTm

A Creative Problem Solving Math Book

Merry X-W/g

25! has ?? digits.

CHRISTMAS is a prime crime time!

God's Math
1 Cross
+ 3 nails
= 4 given

$\sqrt{\text{Xtian}}$
RADICAL
CHRISTIAN

*<|:{>
Santa

2000 Christmases ago, a Son was born...

$f(\text{satan}) = \text{santa}$

Big Bang! Life at 10^{-25} s?

 "I dream of a zillion **angels** worshipping the **King** of kings."

* I'd prove Santa exists statistically! *

Don't open until Dec. 25!

How many wise men visited Jesus? $2 < N \leq 20$

Xmas in 2025 is a Thursday!

© $\hbar\tau 15 \dagger \mho @ 5$

MATHPLUS
PUBLISHING
Specialists in Mathematics Education

Yan Kow Cheong

MATHPLUS Publishing
Blk 639 Woodlands Ring Road
#02-35 Singapore 730639

E-mail: publisher@mathpluspublishing.com
Website: www.mathpluspublishing.com

National Library Board, Singapore Cataloguing-in-Publication Data

Yan, Kow Cheong.

    Christmaths : a creative problem solving math book / Yan Kow Cheong. –
    Singapore : MathPlus Pub., 2011.
    p. cm.
    Includes bibliographical references.
    ISBN : 978-981-08-7655-5 (pbk.)

    1. Mathematical recreations. 2. Christmas. I. Title.

QA95
793.74 -- dc22
OCN690764869

Printed in the United States of America

In memory of my mother

How Ah Moye

(1936–2009)

Preface

In 2003 it dawned on me that an article on "The Mathematics of Christmas" would be apt for the December issue of *Young Generation* (*YG*), Singapore's leading bilingual children's magazine. While researching materials for the feature I realized that a fair bit could be written on the apparently unconnected topics of Christmas and Mathematics.

Following three Christmas articles in December 2003, I went on to write *Joyeux Christmas* in December 2004. Since then I thought that if time permits, I would want to write a recreational-and-problem-solving math booklet, entitled **CHRISTmaths**, which attempts to bring together the joy (or spirit) of Christmas and the spirit (or joy) of mathematics.

During the 2005 festivities, the thought of publishing **CHRISTmaths** crossed my mind again, and after Christmas 2006, I started listing possible topics linking Mathematics and Christmas—what the queen of sciences and the king of public holidays have in common.

My aim for **CHRISTmaths** is that it should appeal not just to a Christmas or Christian audience but to any problem solver who simply enjoys doing mathematics recreationally. In fact, there is nothing special or magical about the number 25 *per se*; one could have favored some other numbers to pose similar nonroutine questions discussed here.

CHRISTmaths should benefit three main groups of readers:

- students numbed by hundreds of sterile drill-and-kill questions found in dozens of Singapore boring assessment titles;
- mathletes who long for some creative problem solving to tickle their mathematical bones;
- problem solvers who long for some intellectual kick out of solving nonroutine questions.

My hope is that **CHRISTmaths** readers would enjoy many hours of intellectual enrichment, while renewing or rekindling their love of recreational mathematics—to experience the *Ah*, *Aha!* and *Ha Ha* of Mathematics.

CHRISTmaths could not have been written without the guidance of the Holy Spirit, and the unmerited favor of our Lord and Savior Jesus Christ. All Praise and Glory be to Him.

K C Yan
@MathPlus

Contents

Biodata of 25

1. I am a square and the sum of two squares:
$$25 = 3^2 + 4^2 = 5^2$$
In fact, I am the smallest square that can be written as a sum of two squares.

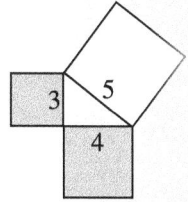

2. I am the hypotenuse of a primitive Pythagorean triple:

$$25^2 = 7^2 + 24^2$$

I can also be represented as Pythagorean triples:

$$25^2 = 15^2 + 20^2$$
$$25^2 + 312^2 = 313^2$$

Moreover, I am the product of the difference of two squares:

$$25 = 13^2 - 12^2 = (13 + 12)(13 - 12)$$

In $25^2 = 625$ and $16^2 = 256$, I pop up at the end and beginning of the output.

3. I can be expressed as the sum of five consecutive old numbers:

$$25 = 1 + 3 + 5 + 7 + 9 = 5^2$$

4. I reveal myself as a sum of five consecutive integers:

$$25 = 3 + 4 + 5 + 6 + 7$$

5. **I am the sum of two triangular numbers: 25** = 10 + 15

6. I am present in this pattern:

   $1 + 2 + 3 + 4 + 5 + 4 + 3 + 2 + 1 = \mathbf{25}$

7. All my powers end in the same digits: 25

   $$25^1 = 25, 25^2 = \mathbf{625}, 25^3 = 15{,}\mathbf{625},$$
   $$25^4 = 390{,}\mathbf{625}, 25^5 = 9{,}765{,}\mathbf{625}, \ldots$$

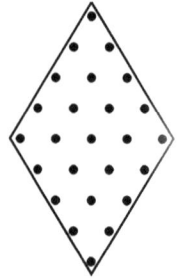

8. Since **25** = 4! + 1, I represent the only solution to $(n - 1)! + 1 = n^k$.   [Liouville]

   When $n = 5$, $(5 - 1)! + 1 = 5^2$

9. **I am the only square that is 2 less than a cube:**

   $$5^2 = 3^3 - 2$$

10. For married couples who have remained faithful for 25 years, I entitle them to a silver wedding. And for a king or queen who has been to the throne for 25 years, he or she can celebrate the Silver Jubilee.

11. In the coins family, I am called a *quarter*.

    I'm a 'whole' – four quarters, not a quarter!

    (25¢)

12. I am a *centered octahedral number*, since

$$\frac{1}{3}(2n - 1)(2n^2 - 2n + 3) = \mathbf{25} \text{ when } n = 3$$

$$n = 3, \frac{1}{3}(2 \times 3 - 1)(2 \times 3^2 - 2 \times 3 \div 3) = \frac{1}{3} \times 5 \times 15 = 25$$

> Centered octahedral numbers represent a crystal ball sequence for cubic lattice. The first few values are $1, 7, 25, 63, 129, \dots$. These numbers have the generating function
>
> $$f(x) = \frac{(1 + x)^3}{(1 - x)^4} = 1 + 7x + 25x^2 + 63x^3 + 192x^4 + \dots$$

13. I am also a *centered octagonal number* (a centered figurate number that represents an octagon with a dot in the center and all other dots surrounding the center in successive octagonal layers.)

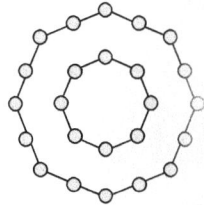

25 dots

I am given by the formula $8\,T_{n-1} + 1$, where T is a regular triangular number, with $n = 3$.

Simplifying, $8\,T_{n-1} + 1 = (2n - 1)^2$, and when $n = 3$, $(2n - 1)^2 = \mathbf{25}$

14. If 6 is added to, or subtracted from, me, I can produce two primes:

$$\mathbf{25} - 6 = 19 \text{ and } \mathbf{25} + 6 = 31$$

15. If 12 is added to, or subtracted from, me, I again yield two prime numbers:

$$25 - 12 = 13 \text{ and } 25 + 12 = 37$$

16. I am the smallest two-digit Friedman number, in honor of Erich Friedman of Stetson University in Deland, Florida.

   In the expression $5^2 = 25$, I use each of my digits precisely once.

   Conjecture: All powers of 5 are Friedman numbers.

17. I am a *Cullen number* since I am of the form $n \cdot 2^n + 1$ (written C_n), with $n = 3$.

18. In number theory, I am the seventh *Hilbert number* since I am of the form $4n + 1$, with $n = 6$:

$$25 = 4 \times 6 + 1$$

   I'm the atomic number of manganese.

19. I am the next favorite term in the sequence:

$$1, 4, 9, 16, \_\_\_\_\_$$

20. I am the only odd perfect square, not equal to 1, which is not the sum of three perfect squares (not equal to 0.)

21. I disguise myself as the number of terminal zeros in five products:

$$1 \times 2 \times 3 \times \cdots \times n, \text{ where } n = 105, 106, 107, 108, 109.$$

22. I am an *automorphic number* because my square ends in 25. In fact, I am the smallest two-digit automorphic number. Two of my friends are 76 and 625.

$$25^2 = 625$$
$$76^2 = 5776$$
$$625^2 = 390{,}625$$

   We're related somehow!

   25 **625** **76**

   I am divisible by 25 if and only if I end in 25, 50, 75, or 00.

23. I associate myself to the Hindu word, Pachisi, for **25**, which is one of the national games of India. And, in Ireland, I am also linked to their national card game 25.

A standard Bingo card has 25 spaces, 24 of which are occupied by numbers.

24. I represent the number of prime numbers less than 100.

25. Last but not least, I am part of a *palindromic equation*:

$$2^5 - (5 + 2) = 25 = 5^2$$

B.C. and A.D.

Here lies **Nerdo**
π 😊 ∞ Σ $\aleph$ φ
Born 3 B.C.   *Died 4 A.D.*

How old was Nerdo?

Look at the tombstone on the left. How old was Nerdo when he died? Most novice problem-solvers would incorrectly give the age of Nerdo as 7 instead of 6 years old.

One common error is the belief that the Christian era started on January 1, 0 (year 0). As a result, January 1, 1 would denote a duration of "1 year old," January 1, 100 a period of "100 years old," and so on.

The trouble with this is that there was never a year 0. The year before 1 A.D. was 1 B.C. The abbreviation B.C. stands for "before Christ," and A.D. stands for "anno Domini" ('in the year of the Lord').

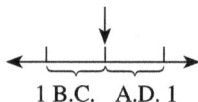

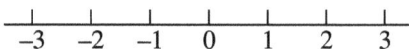

B.C.: 'before Christ'

A.D.: 'anno Domini' ('in the year of the Lord')

$$\xleftarrow{\quad\underbrace{\;|\quad\;|\;}\quad|\quad} \longrightarrow$$
1 B.C.   A.D. 1

$$\xleftarrow{\;|\quad|\quad|\quad|\quad|\quad|\quad|\;}\rightarrow$$
−3   −2   −1   0   1   2   3

The real number line

The real number line has a zero; numbers on the right of zero are the *positive numbers* (1, 2, 3, …), and numbers on the left of zero are the *negative numbers* (−1, −2, −3, …). However, our calendar number goes from 1 B.C. to 1 A.D. without a zero. Jesus Christ was arguably born in 1 A.D., and the year before his birth was 1 B.C. So A.D. 0 and 0 B.C. do not exist.

> A.D. 0 and 0 B.C. do not exist.

Our present-day calendar takes the year Jesus Christ was born as A.D. 1. We express years before his birth in B.C., and years after his birth in A.D. For example, the year 2010 means that it has been 2010 years since Jesus Christ was born. Unless we are comparing years before the birth of Christ, we usually do not use the abbreviation A.D. (or AD) to denote years from 1 A.D.

> Even your birthday is dated by His Birthday!

If a baby has not lived a year, is she *zero* years old?

Back to the tombstone question, from 3 B.C. to 4 A.D., there is only 6 years, rather than 7, because there is no zero year.

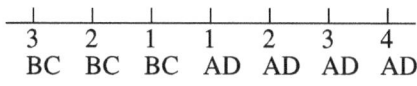

| 3 | 2 | 1 | 1 | 2 | 3 | 4 |
| BC | BC | BC | AD | AD | AD | AD |

When Was Jesus Born?

Jesus Christ was 2000 years old in the year 1997!

Although there is no clue from the Holy Scriptures when Jesus Christ was born, based on our present-day calendar, most scholars believed that it is out by several years. Paradoxically, Jesus is believed to be born some four years "before Christ" (B.C.)!

The sixth-century Roman Abbot and astronomer Dionysius Exiguus (c. 470 – c. 544) was responsible for fixing the birth of Jesus Christ (the "Son of Man") on 25 December A.D. 1. He divided history into "before Christ" and "anno Domini." Technically speaking, Jesus was born in the year 4 B.C.

The year 1 B.C. was the year before the birth of Christ.

The year 1 A.D. began with the birth of Christ.

"A little knowledge about God can be dangerous to your eternal health."
Patrick Morley

Today, in the name of political correctness (PC), non-theists or atheists prefer a dating system that uses BCE ("before the Common Era") and CE ("Common Era"), for they argue that both B.C. and A.D. reflect a Christ-centered viewpoint. So the days of seeing B.C. and A.D. in future may be numbered, as Christian agnostics or atheists push forward their non-religious views in the public sphere.

Your religious position will determine whether you are comfortable with the usage of B.C. and A.D. Meanwhile, let not any PC talk prevent us from the joy of mathematical problem solving, when it comes to solving problems on B.C. and A.D.

Did you know...

A 60-year-old person has lived a full three percent of the time since Jesus was born.

Puzzles on Zero Year

Because there is no zero year in the calendar, many puzzles have been designed to take advantage of this calendrical oddity.

Here are some "B.C./A.D." questions which directly or indirectly relate to the zero year. How many of them can you solve?

0. A Rare Coin

An antique collector was told of a coin that was unearthed in China with the inscription (in Latin): 53 B.C.

"This coin is definitely fake," the collector said without a second thought.

How did he know, without seeing the coin or even a picture of it?

1. The Rise and Fall of the Roman Empire

How many years had elapsed from the rise of the Roman Empire in 753 B.C. to the fall of the Roman Empire in 422 A.D.?

2. **How Many Birthdays?**

   Job is 5 years old, Ruth is 4 years old, and
   Paul is 8 years old. How many birthdays have
   all these children had?

3. **Was She a Centenarian?**

   If a woman was born on the first day of the year 50 B.C. and died
   on the first day of the year 50 A.D., how long did she live?

4. **A "Priceless" Coin**

   At a coin dealer's shop, a shoplifter stole the oldest coin, dated
   156 B.C. If a rare coin is worth $20 for each year before Christ
   that it was minted, how much could he sell it for?

5. **The Most-Lettered Year**

   Which year from the birth of Christ to the
   present requires the most letters when
   written as a Roman numeral?

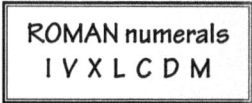

   ROMAN numerals
   I V X L C D M

6. **Periodically Speaking**

   Find the number of years between 34 B.C. and A.D. 60.

7. **Birth of Cleopatra**

   Boadicea died 129 years after Cleopatra was born. Their
   combined ages were 100 years. Cleopatra died in 30 B.C.
   When was Boadicea born?

8. **An Impossibility Made Possible**

   A man was born in the year 1180 and yet died in the year 1163.
   Using the normal calendar, how is this possible?

9. **Death of Thales**

   Thales, a Greek mathematician, was born in 640 B.C. and lived
   for 86 years. When did Thales die?

10. Birth of Plato

Plato, a Greek philosopher and mathematician, died in 347 B.C. after living for 80 years. When was he born?

11. Lifespan of Archimedes

Archimedes, a Greek mathematician, was born in 287 B.C. and died in 212 B.C. For how long did he live?

12. Birth of the Demigod

Sir Isaac Newton, the British mathematician and scientist, died in A.D. 1727 after living for 85 years. When was he born?

13. Triskaidekaphobia (13) & Paraskavedekatriaphobia (Fri 13)

If Christmas 1913 A.D. fell on a Thursday, how many Fridays the 13ths were there in that year?

14. An Iceman's Birthday

The iceman found recently in a glacier was at first thought to be about 5300 years old. Roughly when did he have his 17th birthday?

A. 2700 B.C.   B. 3300 B.C.   C. 3700 B.C.   D. 5300 B.C.
E. 7300 B.C.

15. The Myth of Reincarnation

Using the concept of B.C. and A.D., if we take A.D. 2000 as Year 1 and the year before as Year −1, then in which year were you born? In which years were your parents born?

16. Historically Speaking

Using the concept of B.C. and A.D., answer the following questions with reference to history books.

(a) The Republic of Singapore was founded in 1965. If we take this year as Year 1 and the year before as Year −1, then in which year was the Second World War?

(b) If we take the year the Malays set foot in Singapore as Year 1, then what year is it now?

17. First Day of the Millennium

What was the first day of the 20th century?

18. First Date of the Millennium

What date does the third millennium start?

From *birth date to death date*

LIFE IS SHORT!

1870 – 1920

The length of the dash between the dates on a tombstone is short—life is short.

Where will we go?

1870 – 1920?

What you do with the dash determines where you will go.

Are you Christmas-literate?

Birthmas gift   The double gift for a person unfortunate enough to be born on or around Christmas.

Christmas bomb   When someone wraps another's present in an obscene amount of layers of wrapping paper or other wrapping material, and forms a ball-like shape.

Christmacrastinating   This is what you do instead of writing your Christmas cards, finishing your Christmas shopping, or baking cookies for all your friends, co-workers, and relatives.

Chrismas-casual   A formal yet comfortable style of dress, as for a Christmas party or when you are going to meet the parents.

Christmas clubs   Bank accounts in which customers could deposit a little each week so they could afford presents come Christmas. Around December 13, the banks would disburse the savings plus some interests to participants.

Christmas creep

A phenomenon where the Christmas season starts earlier and earlier each year.

The Xmas season will start in June this year!

Christmas cringe

The feeling of sudden and impending doom after receiving a gift from a co-worker or classmate in spite of the fact that you have nothing in common except that you are co-workers or classmates. This gift is always generic, pointless, useless, and frequently related to some sort of posh lust fad.

Christmas gear

The new clothes that people received for the holidays.

Christmaphobe

Someone who hates and is scared of anything Christmas-related: no un-wrapping presents, singing carols, or Xmas specials on TV, and the like.

Christmaslessness

The state of feeling bored, gloomy, depressed, nostalgic, etc. after Christmas. Feelings are that of having to wait another year for Christmas, which can result in frustration, discouragement, cheerless demeanor, and even despair.

Christmasize

To decorate an item or gift, to look all shiny, bright, and christmassy.

Christmastian A subset of Christians who read the Bible from Genesis to the nativity, and then skip to Revelations at the end, the point being to purposefully exclude the actual teachings of Jesus Christ.

The Christmastian belief system is founded primarily on ancient pagan celebrations that were adopted by early Christians. Christmastians reject peace, love and forgiveness and embrace vengeance, violence, and never-ending war.

Chrismess The aftermath of Christmas celebrations.

For more Christmas words or phrases, click on www.urbandictionary.com

The 12 Puzzles of Christmas

· ·

It's Christmastime, children. Let's have some fun solving some *CHRISTmaths* puzzles. Here are 12 Christmas crackers to whet your mathematical appetite.

1. A Christmas Picture

Using only 4 lines and without lifting your pen or pencil from the paper or retracing lines, connect the 9 dots.

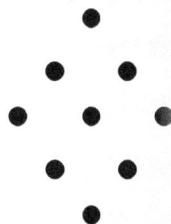

2. Santa Claus Is in Town!

Santa Claus and 4 angels can deliver 5 toys in 5 minutes. How more angels does Santa need to deliver 100 toys in 100 minutes?

3. December 25, 2025

On what day of the week does Christmas Day fall in the year 2025?

4. A Christmas Greetings

Decipher the Christmas message, by filling in the missing letters.

WE WISH YOU A MERRY _ _ _ _.

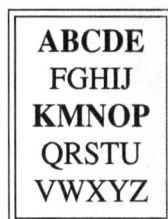

ABCDE
FGHIJ
KMNOP
QRSTU
VWXYZ

*5. The Star of David

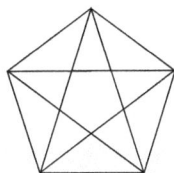

How many triangles you can find in the figure displaying the Star of David?

6. Merry Xmas & Happy New Year

If the value of MERRY CHRISTMAS is 189, the value of A HAPPY NEW YEAR is _____.

7. Crissmass or Krismaths?

"Ther are ate leters im CHRISTMAS."

How many mistakes are there?

8. An Inverted Christmas Tree

Move only 3 stars to get the Christmas tree upside down.

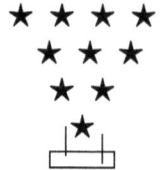

★ ★ ★ ★
★ ★ ★
★ ★
★

9. An Eternal Message

Find the hidden message:

$$\left(\frac{KCOHIR}{IOK}\right)\left(\frac{FIVSPT}{PVF}\right) = \left(\frac{LXORVGE}{XRG}\right)$$

Hint: Simplify.

10. A Christmas Cross

✚ ✚ ✚ ✚ ✚
✚ ✚ ✚ ✚ ✚
✚ ✚ ✚ ✚ ✚
✚ ✚ ✚ ✚ ✚
✚ ✚ ✚ ✚ ✚

Connect 12 crosses with one unbroken line to form a Christmas cross that leaves 5 crosses on the inside and 8 crosses on the outside.

11. The Dotted Cross

How many squares can you create in this dotted cross by correcting any four dots? Note: The corners of the square must lie upon a grid dot.

```
      •  •
   •  •  •  •
   •  •  •  •
      •  •
```

12. The Christmas Gifts

You hear the Christmas love song that goes like:

On the 1st day of Christmas   …
My true love sent to me
A partridge in a pear tree.

On the 2nd day of Christmas   …
My true love sent to me
2 turtledoves, and
a partridge in a pear tree.

On the 3rd day of Christmas   …
My true love sent to me
3 French hens,
2 turtledoves and
a partridge in a pear tree.

…   …

On the 12th day of Christmas
My true love sent to me
12 drummers drumming,
11 pipers piping,
10 lords a-leaping,
9 ladies dancing,
8 maids a-miling,
7 swans a-swimming,
6 geese a-laying,
5 gold rings,
4 calling birds,
3 French hens,
2 turtledoves, and
a partridge in a pear tree.

On the 11th day of Christmas
My true love sent to me
11 pipers piping,
10 lords a-leaping,
9 …

How many gifts were given in all 12 days?

Santa's Itinerary

One type of problems that often comes out in math contests and competitions involves finding the number of ways one can move from one point to another point, subject to some restrictions or conditions.

In a neighborhood, there are 25 houses, how many different routes can Santa take from point X to point Y, moving only upwards and to the right?

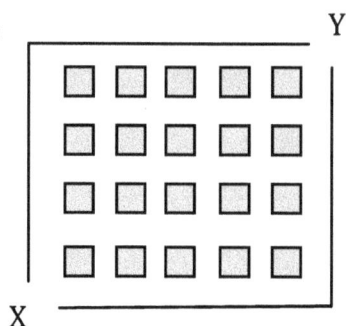

Let's look at some simpler cases before attempting the above problem.

Worked Example

Any path from A to B must follow the horizontal or vertical lines of the grid. For each of the grids given below, find
(a) the length of the shortest path from A to B,
(b) the number of shortest paths from A to B.

(1)

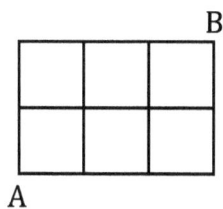

(2)

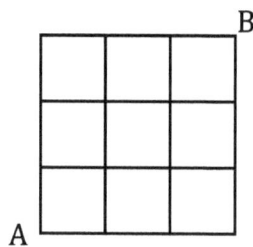

(3)

(a) The length of the shortest path is the sum of units of the width and length of the grid.

For figure (1), the shortest path is 3 units.
For figure (2), the shortest path is 5 units.
For figure (3), the shortest path is 6 units.

(b)

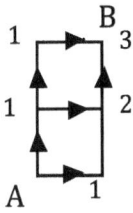

3 ways

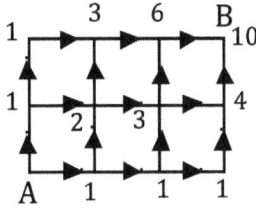

10 ways

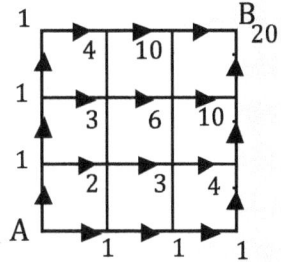

20 ways

Note that there is a connection between the number of shortest paths and Pascal's Triangle.

Answer for grid (1): 3, 3 occur in
row 3: 1 │3│ │3│ 1

Answer for grid (2): 5, 10 occur in
row 5: 1 │5│ │10│ 10 5 1

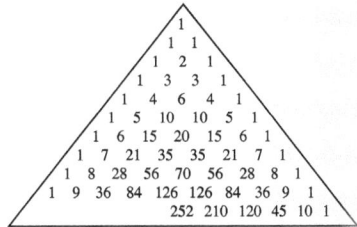

Answer for grid (3): 6, 20 occur
in row 6: 1 │6│ 15 │20│ 15 6 1

In general, given width w units and length n units, the number of shortest paths is the wth term (or nth term) in the $(w + n)$th row.

For example, in grid (3), the number of paths found is the 3rd term in the $(3 + 3)$th, or 6th row.

What is the number of shortest paths for the grid whose lengths are 9 units and width 8 units? Either use Pascal's Triangle up to 17 rows and using the above result. Or use the factorial notation.

The wth term in the $(w + n)$th row of Pascal's Triangle is given by $^{w+n}C_w = \frac{(w + n)!}{w!n!}$, so the number of paths on a 9×8 grid is $^{17}C_8 = \frac{17!}{8!9!}$ = 24,310.

Practice

1. If Santa were to move from point X to point Y. The path between X and Y are such that any distance must be 4 cm. How many possible routes can Santa take from X to Y?

    X
    1 cm
    1 cm
    1 cm 1 cm Y

2. Joseph wants to walk all the different routes from his house to his school. The map below shows all the streets between Joseph's house and his school.

    school

    How many different ways can he walk to school?

3. In Euclid's Homes, there are 16 houses. How many different routes can you take from point X to point Y, moving only upwards and to the right?

    Y

    X

4. If there are 25 houses, how many different routes can one take from point X to point Y, moving only northward and eastward?

5.

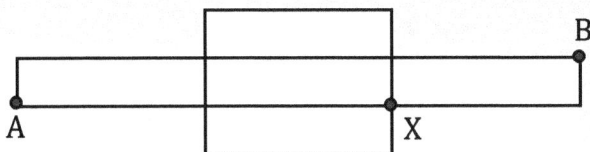

Without passing through any point in the network more than once, how many different paths are there from point A to point B via X?

6. In how many different ways can Father Noel walk from *A* to *B*, always moving north or east, without passing through *X* and *Y*?

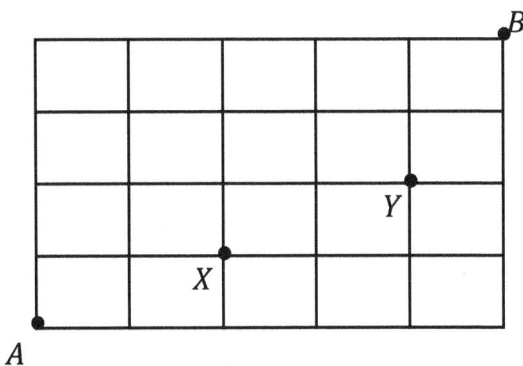

7. The figure shows a rectangular hall with 40 rooms. Every two adjacent rooms are connected, and one can walk from one room to another in an eastward or southward direction. In how many ways can Father Claus walk from room X to room Y passing room P and room Q?

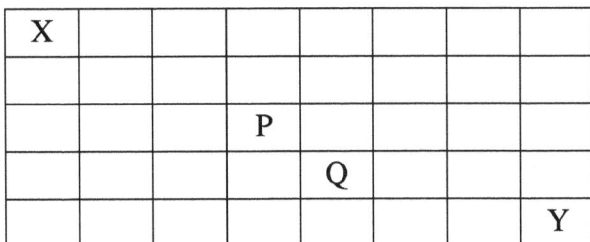

X							
		P					
			Q				
						Y	

8. The diagram shows that the street plan of a town. The police are chasing a serial killer who is making his way from A to B. Policemen are stationed from A so that every possible route from A to B is covered to ensure that the killer will be caught, as the police can move either northwards or eastwards, how many policemen must be mobilized?

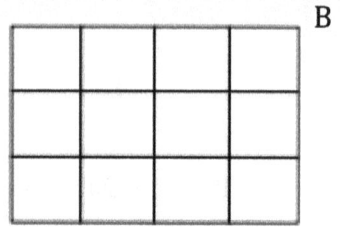

9. If you can only travel northward and eastward, how many different routes are there from A to B passing through XY?

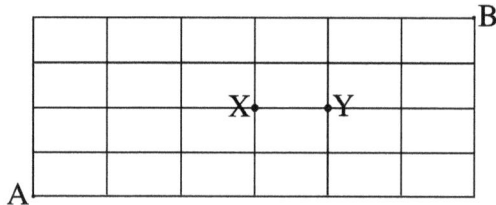

10. In the lower half of the figure, Noël can only move up and to the right, while in the upper half, he can only move up and to the left. How many possible routes can he move from X to Y?

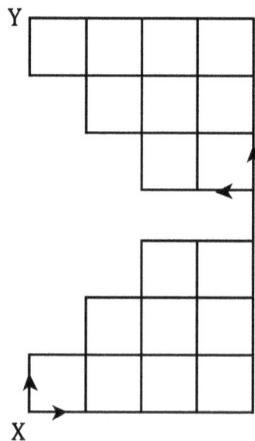

11. Pilate starts at point X passing through a number of blocks to reach point Y. If he were to travel the shortest path to his destination, how many different routes can he take?

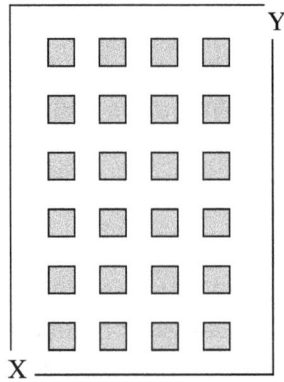

12. If Emmanuel were to follow the arrowed routes in the diagram, how many different routes could he travel from P to Q?

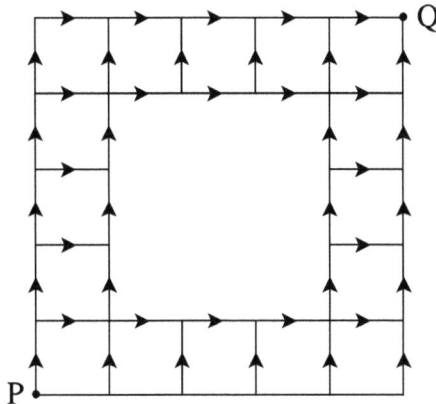

12 Daffynitions of CHRISTMAS

Daffynitions are toss-off definitions, referring to those rib-nudging, brain-tickling one-liners. Let's look at those infuriating, "anti-Websterisms" when applied to Christmas.

Christmas 1. A time of the year when people empty their bank accounts, come to blows with their colleagues and friends and suffer embarrassment at office parties, but which is nonetheless referred to as a period of rejoicing.

2. The season to be jolly miserable and financially unstable.

3. A feeling of a 25-hour day with no fewer than 25 disappointments.

4. A time you buy this year's presents with next year's money.

5. A time when you spend more, give more, party more just to avoid being left out.

Christmas: A day set apart and consecrated to gluttony, drunkenness, maudlin sentiment, gift-taking, public dullness and domestic behaviour.

Ambrose Bierce,
The Devil's Dictionary

6. A feeling of loneliness in the midst of crowds.

7. A time when you are insincerely generous, buy things hardly everyone wants and give them to those you don't like.

8. **The time of the year when obscene amounts of money are wasted, in getting things that nobody wants, and nobody cares for after they have got them—the vulgar commercialization of Christmas.**

9. A day of the year when the crime rate is at its prime—the have-not's want to get even with the have's.

10. **A day when the birthday of Jesus Christ is relegated to the Xmas Eve's service, with everyone wishing each other "Merry Xmas" instead of singing "Happy Birthday, Jesus."**

11. An annual commercial carnival that celebrates the birth of the Shopping Center, usually lasting about three to four months, and starting earlier every year, with Christmas 2025 predicted to usher in August.

12. **The time of year when Santa rises from the dead, hypnotizes half a dozen reindeer hanging out in a log cabin to be his slaves and uses them to carry his sleighful of China-made toys to millions of children's houses, before slaughtering the overstressed animals and storing them in the freezer.**

My *daffynition* of
CHRISTMAS is
... less math ...

A Christmas Spell

. . . .

Counting the number of ways a given word may be counted based on certain conditions is not an uncommon question in a number of junior mathematical olympiads. Let's look at a few of them.

Worked Example 1

In how many ways can the word CHRISTMAS be spelled from the top and working down?

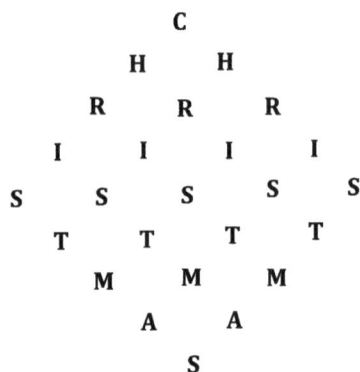

```
                    C
              H           H
          R       R       R
      I       I       I       I
   S      S       S      S       S
      T       T       T       T
          M       M       M
              A       A
                  S
```

Solution

The number of ways of spelling the word CHRISTMAS from the top and working down is 70.

If we can select any letter from each row, then the number of ways of forming CHRISTMAS is

$1 \times 2 \times 3 \times 4 \times 5 \times 4 \times 3 \times 2 \times 1$
$= 2880$.

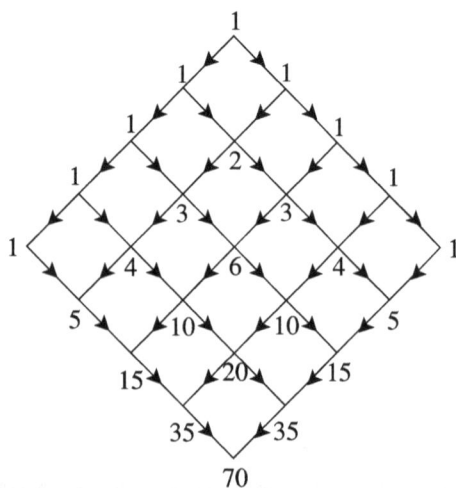

A Math Spell

In how many ways can the word MATHEMATICS be spelled starting from the top and working down through the array if:

(a) we can select any letter from each row,
(b) we can only select either one of the two neighboring letters directly underneath?

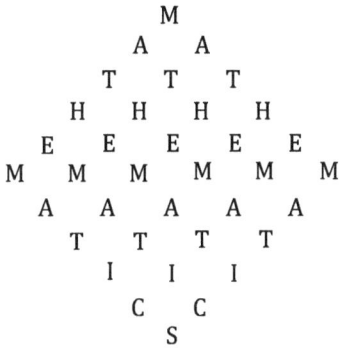

```
              M
           A     A
        T     T     T
     H     H     H     H
   E     E     E     E     E
 M     M     M     M     M     M
   A     A     A     A     A
     T     T     T     T
        I     I     I
           C     C
              S
```

> Think about
> the Pascal's
> triangle.

In how ways can you get to each of the A's in the second row?

Solution

(a) $1 \times 2 \times 3 \times 4 \times 5 \times 6 \times 5 \times 4 \times 3 \times 2 \times 1 = 86{,}400$ ways

(b) The number of routes to each letter is given by:

```
              1
           1     1
        1     2     1
     1     3     3     1
   1     4     6     4     1
 1     5    10    10     5     1
   6    15    20    15     6
     21    35    35    21
        56    70    56
          126   126
              252
```

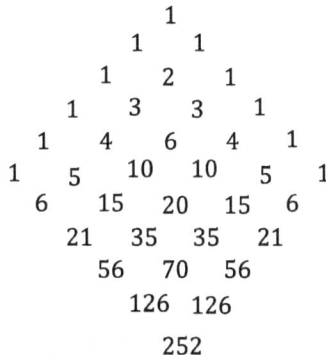

Hence the number of ways is 252.

Counting MATH's

How many MATH's can you find?

```
                    H
              H     T     H
        H     T     A     T     H
  H     T     A     M     A     T     H
              T     A     T
        H                       H
              T
        H           H
              H
```

Solution

Number of MATH'S
= 8 × 3 + 4 × 1
= 24 + 4
= 28

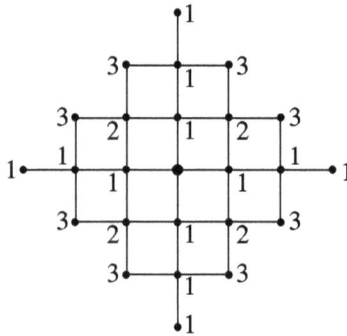

Practice

1. (a) How many ways can you spell CHRISTMAS in this
         diagram?

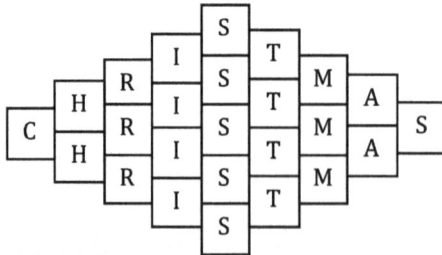

(b)

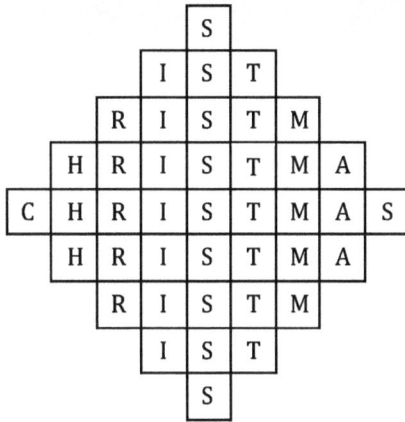

How would you show that you did not under-count or over-count?

Are the answers to (a) and (b) related?

Hint: Use shorter words.

2. In how many different ways can the word THINKER be read from left to right in this diagram, with each letter being selected from each row?

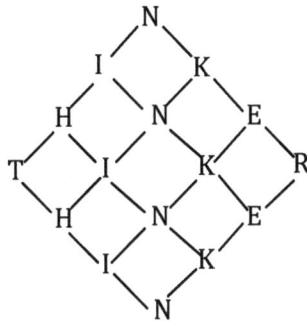

3. In how many ways could THINKER be spelled if you do not have to go to the next letter along a line—for example you could go from either of the H's to any of the I's?

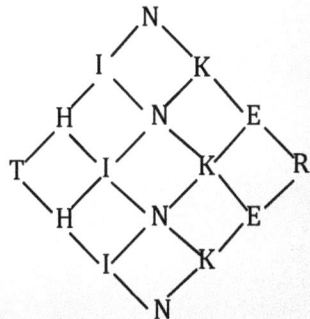

4. How many ways can a bee
   travel from cell C to cell T in
   the beehive below in
   (a) three steps,
   (b) four steps,
   (c) five steps?

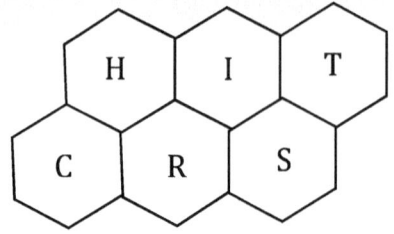

5. In how many ways can you form the word "SANTA" from the
   diagram below?

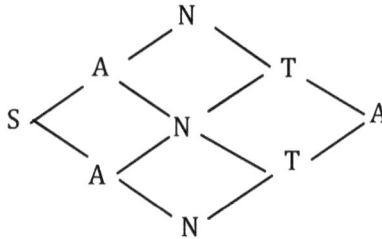

6. In how many ways can you form the word FORGIVE be read in
   the diagram below?

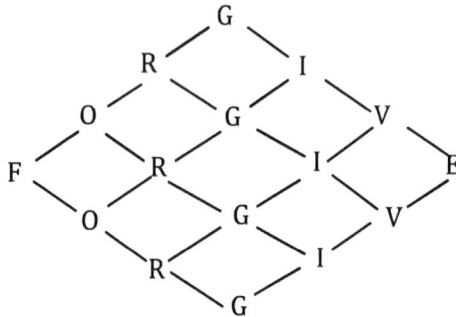

Guesstimation on Xmas Day

Estimation is often confused with approximation, which is about rounding off or truncating numbers, based on some well-defined rules. Although estimation and approximation are used interchangeably in daily conversation, however, these two terms have different meanings in mathematics.

Estimation tends to be wrongly associated with wild guessing and taken to mean approximating large or small numbers. While it involves approximation, estimation goes beyond mere educated or intelligent guessing of events in real life, whereby the required data may not easily be obtainable or measurable, or the process of finding them is simply impractical or impossible, due to physical constraints, costs, or time.

To be a good *guesstimator*, or to make sense of often confusing and sometimes contradictory numbers used in the media, you need only to:

Guesstimate is a portmanteau word combining "guess" and "estimate."

(1) understand what large or small numbers mean, and
(2) make rough, common-sense, estimates starting from just a few basic facts.

Back-of-the-envelope Math

The process of guesstimation is often likened to a quick-and-dirty exercise which could be done at the back of a paper napkin or envelope while waiting for the waitress to bring up your order.

Be a Fermi Disciple!

Today, a number of companies use guesstimation questions in job interviews to judge the intelligence and flexibility of their applicants. Leading computer and consultancy firms and investment banks, such as Microsoft and Goldman Sachs, may ask questions such as *How big is the market for Christmas presents worldwide? How many Rubik cubes could fill a Boeing? How many kilograms of toys are delivered by Father Santa on Christmas Eve?*

Multinationals use these questions to test the applicants' abilities to think on their feet and to apply their mathematical skills to real-world problems. These problems are frequently christened "Fermi problems," named after the legendary physicist Enrico Fermi (1901–1954), who delighted in creating and solving them.

Like cycling and swimming, guesstimating is a lifelong skill which will help you to estimate almost anything, from the earnings by retailers during the festive season to the number of unhappy people on Christmas Eve to the number of partygoers who oversleep on Boxing Day. Are you ready to put on your guesstimation cap? Compare your answers to the questions below.

Example 1

CHRIST*maths for the Blind* has about 252,525 letters in it. Estimate the thickness of the pop math book.

> Mandatory Parental guidance
>
> CHRISTmaths
> for
> the Blind
>
> K C Yan

Solution

The book has 252,525 letters.

Taking a word to be on average 5 letters long, the book will then have about

$$\frac{252,525}{5} = 50,505 \text{ words.}$$

Now, an average page has about 500 words.

So there are about $\frac{50,505}{500} \approx 100$ pages in *CHRISTmaths for the Blind*.

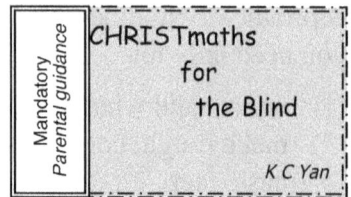

Example 2

> How many airplane flights do Singaporeans take during the festive season?

Solution

Let's begin with the number of Singaporeans, and let's estimate the number of plane flights each of us take during school holidays.

There are about 3,500,000 Singapore citizens. Say, about one fifth of them travel once a year (i.e., two flights) during the December school holidays, and a small percentage of the population (say 10 percent) travel more than that. This means that the number of flights per person per year is between two and four.

Let's take two for an average. Therefore, the total number of flights during the December school holidays is

$$\frac{1}{5} \times 3\,500\,000 \text{ people} \times 2 \text{ flights/person-year}$$
$$= 1,400,000 \text{ passengers/December holidays}$$

Hence there are about one and half millions airplane flights by Singapore citizens every festive season.

Some General Guidelines

To get a reasonably good estimate, there are some general guidelines you can follow. Here are some of these.

1. Go from the abstract to the concrete.

2. **Tap on your everyday experiences.**

3. Show your logic — assume your reader has no prior experience.

4. **Use common sense and readily available data.**

5. Try a few estimates and see whether they are close to each other.

There are many different ways of estimating an answer.

Here are some guesstimation Xmas crackers, which will keep you quantitatively pre-occupied during the festive season.

Practice

1. If January 1900, year 1 was a Monday, what day of the week will it be in 25,252,525 days later?

2. Is the distance to the nearest planet more than 252,525 km?

> The Bible goes metric!

3. Is the distance to the nearest galaxy more than 10×10^{25} km?

4. Can a skyscraper be as high as 252,525 meters?

5. Does a 25-storey building weigh as much as 252,525 tons?

6. Were you born 252,525 minutes ago?

7. How old would you be in years if you were born a billion seconds ago?

8. What is the total length of all the hair on an average woman's head?

9. How much wrapping paper is used during Christmas every year?

10. How long would the rolls of stick tape bought for Christmas be?

11. How old would you be if you were 25 in dog years?

12. How many Christmas cards are sent round the globe every year?

13. How many turkeys are consumed in the world every year during the festive season?

Selected Answers

7. About 32 years old.
8. About 10 km.
9. Christmas wrapping paper comes with about four times round the world.
10. The rolls of stick tape bought for Christmas is more than 240,000 miles or 384,000 km (Earth to Moon).

Thou Shalt Guesstimate More on Xmas Day!

To sharpen your estimation skills, and to groom yourself in becoming an *Estimation Specialist*, here are more opportunities to apply more back-of-napkin calculations.

A *Not So Merry* Christmas

What percent of festive shoppers voted "I Wish It Could be Christmas Every Day" as the most irritating Christmas song?*

The Math of Life and Death

What is the probability or chance of someone being born on Christmas Day and also dying on Christmas Day?

What percent of Xmas babies die on their 25th birthday in a developing country?

What are the chances of a human couple giving birth to a Christmas twin or triplet?

What percent of babies born to Christian parents will be a born-again Christian?

If the probability of one's living up to 125 years old is 0, what would be the probability of one's living up to 25 years old?

What percent of Singapore schoolchildren do not believe in Santa Claus?

How many trees worth of Christmas cards that the average Singapore resident throws away each year?

What is the average number of days Singaporeans spend singing carols every festive season?

---

* According to the authors of *What are the Odds?*, the answer is 33.33 per cent. (Sharpe & Schlaifer, 2007, p. 235)

What percent of people in Singapore do not believe that God exists?

How much money does the Singapore government spend on festive decorations every year?

What are the odds that someone's phone 8-digit number in Singapore will be 25252525?

What are the odds that *CHRISTmaths* (25th edition) will top the bestsellers' list in 2025?

What are the odds that *CHRISTmaths* will be sold continuously for a quarter century?

You are born twice, but you die once.

7 Beautiful Xmas Series

1. The Harmonic Series

$$\frac{1}{1} + \frac{1}{2} + \frac{1}{3} + \cdots + \frac{1}{25} + \cdots \rightarrow \infty$$

2. The Exponential Series

$$1 + \frac{1}{1} + \frac{1}{1\times2} + \frac{1}{1\times2\times3} + \frac{1}{1\times2\times3\times4} + \cdots + \frac{1}{1\times2\times\ldots\times25} + \cdots$$
$$= e$$
$$= 2.7182818284\ldots$$

3. A Logarithm ie Series

$$\frac{1}{1} - \frac{1}{2} + \frac{1}{3} - \frac{1}{4} + \frac{1}{5} - \frac{1}{6} + \frac{1}{7} - \cdots + \frac{1}{25} - \cdots$$
$$= \ln 2$$
$$= 0.6931471805\ldots$$

4. The Liebniz-Gregory Series

$$\frac{1}{1} - \frac{1}{3} + \frac{1}{5} - \frac{1}{7} + \frac{1}{8} - \frac{1}{11} + \frac{1}{13} - \frac{1}{15} + \frac{1}{17} - \frac{1}{19} + \frac{1}{21} - \frac{1}{23} + \frac{1}{25} - \cdots$$
$$= \frac{\pi}{4}$$
$$= 0.78539816\ldots$$

5. The Euler Series

$$\frac{1}{1^2} + \frac{1}{2^2} + \frac{1}{3^2} + \frac{1}{4^2} + \cdots + \frac{1}{25^2} + \cdots = \frac{\pi^2}{6}$$

$$\frac{1}{1^2} + \frac{1}{3^2} + \frac{1}{5^2} + \frac{1}{7^2} + \frac{1}{9^2} + \cdots + \frac{1}{25^2} + \cdots = \frac{\pi^2}{8}$$

$$\frac{1}{1\times2} + \frac{1}{2\times3} + \frac{1}{3\times4} + \frac{1}{4\times5} + \frac{1}{5\times6} + \cdots + \frac{1}{24\times25} + \cdots = 1$$

Can you prove these elegant results?

Santa Claus

12 Challenges @ Christmastime

....

Here are a bagful of Christmas challenges — the Twelve Games of Christmas — prepared by Father Santa and his team of problem posers for you and your family to solve during this festive season.

1. What is the last digit of $25 \times 25 \times 25 \times \cdots \times 25$?

2. Find the angle of each point on the regular five-pointed star.

3. To number the pages of a mathematics almanac, a printer had used 3133 digits. How many pages has the almanac?

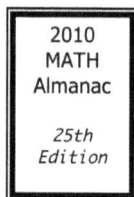

    > 2010
    > MATH
    > Almanac
    >
    > 25th
    > Edition

4. Can Christmas Day fall on any day of the week?

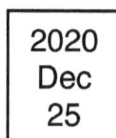

    > 2020
    > Dec
    > 25

5. Using each letter only once, how many mathematical words can be formed from this festive greetings?

    O WE WISH YOU ALL A MERRY CHRISTMAS
    AND A HAPPY NEW YEAR

6. Find the value of

$$\sqrt{25 + \sqrt{25 + \sqrt{25 + \cdots}}}.$$

This nested expression is infinite but it has a finite value.

7. Last Christmas, Ai Ling posted 50 items. Each local card cost 40¢, each letter cost 45¢, and each overseas card cost 95¢. She sent more letters than overseas cards. If she spent a total of $28.00, how many of each item did she send?

8. Find the number of triangles contained in the figure displaying the Star of David.

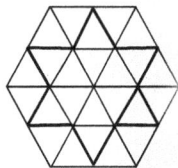

9. Two candles were lighted on Christmas Eve. One was 1 cm longer than the other. The longer candle was lighted at 4:30 and the shorter one was lighted at 6:00. At 8:30 they were the same length. The longer one burned out at 10:30, the shorter one at 10:00. How long was each candle at first?

10. Solve for x: $\sqrt{x + \sqrt{x + \sqrt{x \cdots}}} = 25$.

11. It is late on Christmas Eve and little Noel is waiting for the Christmas tree to be finished. At exactly what time will this happen?

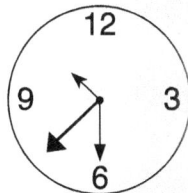

12. In the sequence MERRYXMASMERRYXMASMERRY…, what is the 2525th letter?

A Mathematician's Musings on Xmas Day

· ·

Mathematicians of all stripes are known not to rest on the most popular public holiday of the year—they are crazy busy, musing over odd-sounding cohundrums. Could you help them to answer some of these?

What day of the week was Xmas 1925?

What could the definition of a "Christmas number" be?

Could Jesus Christ have died on Friday 13?

Is God Truly the Great Mathematician?

What if Jesus were born on December 25?

What could "Christening Numbers" be?

What day of the week is Christmas Day most likely to fall? Is there a Christmas pattern?

What is the probability that a Christmas baby will also die on Christmas Day?

What is the probability that a Christmas baby will die at the age of 25?

Could you be the great-great grandchild of Adam and Eve 25 generations ago?

When will the next solar ellipse fall on Christmas Day?

Are there any shapes with fractal dimension 2.5?

What is the sum of the internal angles of a [25,5] star?

Marriageable dates for your descendants:
UK: 25/12/2512; US: 12/25/5221

How many UFO sightings were ever made on December 25?

How many babies are born on Xmas Day every year round the globe?

How many newborn Christmas babies are induced every year?

How many people die on Christmas Day every year?

Born: 25 Dec 1925
Died: 25 Dec 1950

Twelve meta-mathematical proofs that Santa is a myth!

What if we design a 25-hour clock?

Mathematical Graphiti

MERRY CHRISTmaths

The mystery of the Trinity is solved at last – God is working mod 2.

$25 TREES

Don't count sheep. Look for the Shepherd.

If God wanted us to decimalize why were there twelve disciples?

Christmas was sun-worship!

The fØØl hath said in his heart there is nØ empty set.

The Unholy Trinity
Santa Claus
St. Nicholas
Father Christmas

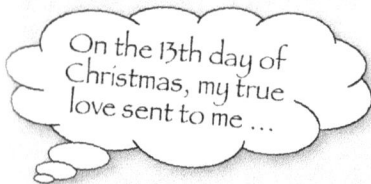

I sin at 45 therefore I am.

Simplify the holidays!

On the 13th day of Christmas, my true love sent to me ...

Father Christmas was hanged at 12:25!

My Philosophy 4 Xmas

Santa Claus Really Exists!

Mathematicians are hard @ work & active on Xmas Day

I don't count, therefore I'm not.

Santa won't be home 4 Xmas!

Xmas Philamath

· ·

> **A sheet of postage stamps has 25 rows with 25 stamps in each row. What is the least number of tears you need to get all the stamps apart?**

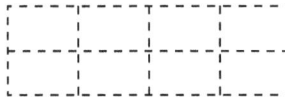

PASCAL   0,82 €

€ 0,65

Pythagoras

Before solving the above problem, let's consider some simpler cases first.

Example 1

A sheet of postage stamps has 2 rows with 4 stamps in each row. What is the least number of tears you need to get all the stamps apart?

Method 1

Simply folding the stamps along the perforation would require no more than four tears, as shown below. Two ways are shown below. Can we do better than that?

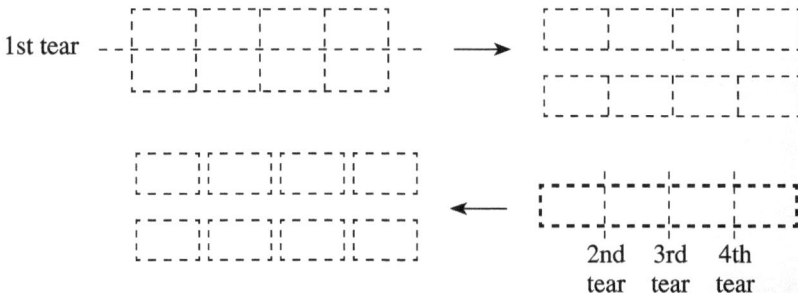

1st tear

2nd   3rd   4th
tear   tear   tear

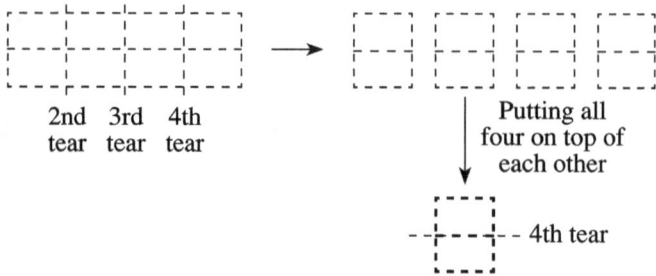

2nd 3rd 4th
tear tear tear

Putting all
four on top of
each other

4th tear

Let's look for a better way of getting all eight stamps apart.

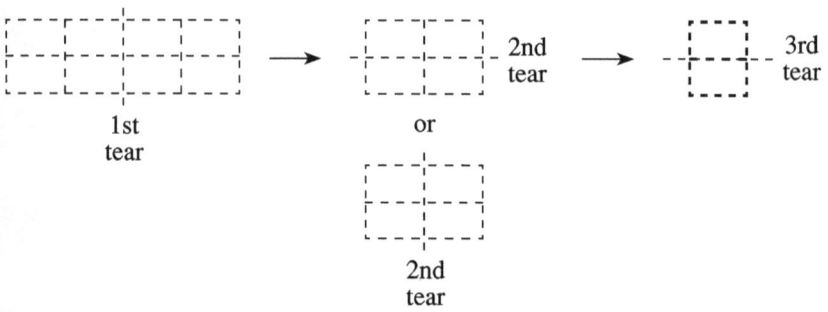

1st
tear

or

2nd
tear

2nd
tear

3rd
tear

So, three tears are sufficient to get all the stamps apart. In fact, there are more than one way to use three tears to get the stamps apart.

Example 2

How many ways can you fold a strip of 3 stamps, if it does not matter whether the stamps are facing up or down?

Solution

Let's label the stamps 1, 2, and 3.

The possible folds are:

$$123 \quad 213 \quad 312$$
$$132 \quad 231 \quad 321$$

Hence there are 6 ways of folding a strip of 3 stamps.

1. A sheet of postage stamps has 3 rows with 4 stamps in each row. What is the least number of tears you need to get all the stamps apart?

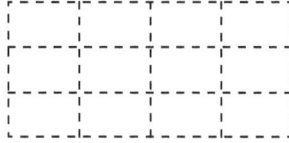

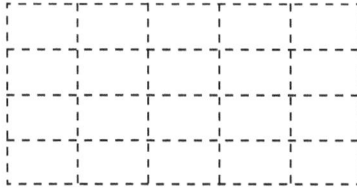

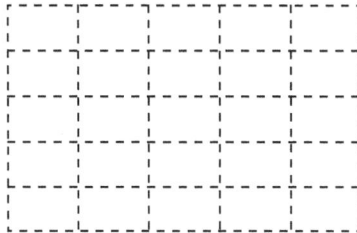

2. A sheet of postage stamps has 4 rows with 5 stamps in each row. What is the least number of tears you need to get all the stamps apart?

3. A sheet of postage stamps has 5 rows with 5 stamps in each row. What is the least number of tears you need to get all the stamps apart?

4. A sheet of postage stamps has 25 rows with 25 stamps in each row. What is the least number of tears you need to get all the stamps apart?

5. You are given a large supply of 10¢ and 20¢ stamps. How many different ways could you stick these stamps on an envelope to send a letter to Father Santa?

    Note: The stamps have to be side by side and the right way up.

For example, if the postage costs 30¢, here are the three possible arrangements.

| 10¢ | 10¢ | 10¢ | | 20¢ | 10¢ | | 10¢ | 20¢ |

How many different ways can you stick stamps worth 40¢, 50¢, and so on?

6. A rectangular sheet is made up of 25 stamps. If folding is not allowed, what is the least number of tears needed to break the sheet into its individual stamps?

7. How many ways can you fold a strip of 4 stamps, if it does not matter whether the stamps are facing up or down?

8. You have an unlimited supply of 2¢ and 5¢ stamps only. Which amounts can be made using just these stamps? Which amounts cannot be made?

9. A post office has stamps of denominations 1 cent, 2 cents, 3 cents, and 4 cents. How many ways can Gill make postage of 4 cents out of these four types of stamps if the order in which the stamps are arranged on an envelope matters?

12 Myths about Christ and Christmas

1. Jesus Christ was born on December 25.

Dec 25

2. **Jesus Christ was born in the year zero.**

25/12/000

?12/0
25/?/?
?/?/?

What if there were 3 women?

3. There were three wise men offering gifts to the Son of Man.

4. **The founder of the Christian religion was called Jesus.**

5. Jesus claimed he was the Son of God.

6. **Xmas is a modern, commercial abbreviation for Christmas.**

7. There is only one biblical account of Jesus' birth.

Adacadabra
~~Algebra~~
on
Xmas

8. **The birthplace of Jesus was in Bethlehem.**

9. Jesus was born when Herod was King of Israel and Quirinius was Governor of Syria.

10. **Jesus was born around the year 7 B.C.**

11. The year 2010 means that 2010 years have gone by since Jesus Christ was born.

12. A.D./B.C. stands for "After Death/Before Christ."

Let's look at some brief explanations to disprove these myths.

Myth 1: Jesus Christ was born on December 25.

No one knows the exact date of Jesus' birth. December 25 was chosen around A.D. 354 to celebrate his birth. It was chosen to take advantage of existing celebrations: Winter Solstice (Dec. 21) and Roman feast day (Dec. 25).

Myth 2: Jesus Christ was born in the year zero.

No one was born in the year zero. There was never such a year because zero was not used in the Roman counting system.

In the 6th century A.D., the monk Dionysius Exiguus (c. A.D. 470 – 544) was asked by Pope John I to determine the dates of Easter for the years 527 to 626.

Using the Rome's founding as a starting point, Dionysius's computations showed that Jesus had been born in the year 753 in the Roman calendar (since the foundation of the city). He established this as the year A.D. 1: "Anno Domini," meaning the first year of Our Lord. It was only in relatively modern times, when negative dates were used (i.e., B.C.: Before Christ) that A.D. was equated with "After Christ."

What's the difference between "year zero" and "zero year"?

A year zero was never introduced, however, so that 1 A.D. followed immediately after 1 B.C. On this basis, the third millennium also began on January 1, 2001, as the 20th century did on January 1, 1901.

Myth 3: There were three wise men offering gifts to the Son of Man.

Tradition tells us that three wise men, or kings, from the east came on camels to pay homage to the Christ-child. The Bible says that "wise men from the east" came, but it does not specify how many, how they got there, or when they visited Jesus. They were probably magi, learned men who studied astrology, who walked rather than rode camels, and they might have arrived up to a year after Jesus' birth, in which case he would no longer be living in the stable, but in a house.

Myth 4: The founder of the Christian religion was called Jesus.

He was called Joshua or Jeshua, which means "He who will save." Jesus is the Greek version of these names. The title "Christ" was first bestowed upon him by his followers after his death. It is also Greek and means "The Anointed," translated from the Hebrew term *masiah* (messiah).

Myth 5: Jesus claimed he was the Son of God.

According to the Gospels, Jesus never did. He referred to God as his father, but he referred to himself as "the Son of Man." However, through his earthly ministries and miracles, the claim that Jesus is the Son of God is indisputably settled, although nonbelievers may think or argue otherwise.

Myth 6: Xmas is a modern, commercial abbreviation for Christmas.

"X" has been used as an abbreviation of the word "Christ" for hundreds of years. It is the first letter of his name, *Christos*, in Greek, written as Xpucroc—it begins with the letter *chi*, or *x*. Xmas has been used as a shorter form of Christmas since about 1755.

Myth 7: There is only one biblical account of Jesus' birth.

Two accounts are in the Bible. In the Luke gospel, Mary and Joseph made their way to Bethlehem, where they found no place at the inn, and the Christ-child was born in the stable.

The Matthew gospel recounts how Mary and Joseph lived in Bethlehem with their small son. Attracted by a star, the Magi came from the east and honored the child, but also brought his birth to the notice of King Herod, and the family was forced to flee to Egypt.

Myth 8: The birthplace of Jesus was in Bethlehem.

Church historians assume that Jesus came into the world perfectly normally in the house of his parents. In their secular view, the stories about Bethlehem arose later because of the Jewish prophecy that this town would be the birthplace of the Messiah. They argued that even if there had been a tax census and Joseph had property in Bethlehem, there would not have been the slightest reason for him to have had to take his pregnant wife with him.

Myth 9: Jesus was born when Herod was King of Israel and Quirinius was Governor of Syria.

The evangelist Luke tells us that Jesus was born during Herod's reign and at the time of the census carried out by Syria's governor, Quirinius. By most accounts, Herod died in 4 B.C., although some say as late as 1 B.C., while the date at which Quirinius took office and the tax census occurred is a full 10 years later, in A.D. 6.

Myth 10: Jesus was born around the year 7 B.C.

To calculate the date of Jesus' birth, many had used the Star of Bethlehem as a starting point. Even Johannes Kepler (1571–1630) who cast a horoscope of Jesus Christ assumed that the Star of Bethlehem was a conjunction of Jupiter and Saturn in the year 7 B.C., which means that both planets appeared so close together that they shone like a single, particularly bright star.

Myth 11: The year 2010 means that 2010 years have gone by since Jesus Christ was born.

The absence of a zero year and the uncertainty arising from conflicting accounts (or sources) on the birth year of Jesus only point to the fact that although the origin of our present-day calendar is based on the birth of Jesus Christ, however, to claim that 2010 years had elapsed since Jesus' birth is arguably questionable, to say the least.

Myth 12: A.D./B.C. stands for "After Death/Before Christ."

Historians claimed that the years B.C. took place before the birth of Christ, as did about five years A.D.—Christ was born around 6 A.D. True, B.C. stands for "Before Christ." The common misconception is that A.D. stands for the "After Death." In fact, A.D. is Latin: *Anno Domoni*, meaning "The Year of Our Lord."

Non-theists or the politically correct (or PC) crowd suggests replacing B.C./A.D. with B.C.E/C.E, abbreviating "Before the Common Era/ Common Era" in an effort to remove specific Christian reference.

Mathematical Graphiti

25 is the new 18!

XMA$ 2010 AUCTION$
by Christie

- The Xmas dress worn by Sharon Stone on Xmas 1973 fetched $26,000 at auction in 2001

- A pair of Xmas dirty stockings sold on eBay for an undisclosed sum – bid won by a fetish Emir

- A Xmas card, handwritten by J. K. Rowling, sold for $12,500 in 2005, for charity!

- Sir Isaac Newton's teeth were sold for $730

The BIBLE CODE

```
HMSSLOAYINMBTGDAVSKSM
KPEJJPURNFLBDCOPGTUSLPS
VKARAIFQIPASNUQNJLPLHKL
FSMKUPSNDULSVPMAFAVBS
UWNYTBIVSPAVNWNLJNMOP
NYMWIFAJOFGWTDIMIKOJK
PWIBMPDNHFLPGNYSBGOPJ
WBBGWFBILMGHARDXLOJL
AWASIKADWVDMMTNMLI
QGWSBKPMGTNLDBJWYBKP
KPTEVJYAFWQMBFSVXZOLP
```

Warning! Jesus may come back on Dec. 25!

$n!$ ends with 25 zeros. What's n?

All taken up!
Santa.com
Santa.org
Santa.net

Santa ⊂ Satan ⊂ Savior

A Look-see proof

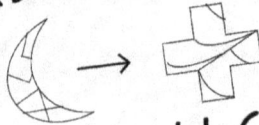

From Crescent to Cross

A Creativity Test ?

How many uses for the number 25 can you think of?

Come up with 25 uses in 25 min.

Mathematical Graphiti

Anagramatically Speaking

Christmas time. = It charms mites.

Christianity = I cry that I sin!
= Charity's in it.

Christian. = Rich saint!
= Rich in sin
= Stain-rich

Ten Commandments
= Can't mend most men

Three scores years and ten.
= So thy career nearest end.

News Reports

CHRISTMAS has been cancelled. They've found the Father.

666 bedevils both believers and non-believers.

A CHRISTmaths Poem

Doing math in a Christ-like spirit
When divinity crosses numeracy
The real joy of mathematics
The true spirit of Christmas

A: Where does Santa Claus come from?

A: Santa is Turkish!

Santa Claus, Spiderman & Superman having a conversation in space

Here We Come A-Counting

On the 13th day of Christmas
My two loves sent to me
One ream of paper, two pencils
Three erasers, four protractors
Five compasses, six set squares
Seven calculators, eight batteries
Nine abacuses, ten math problems

Futuring with 25

- A 25-hour day
- A 25-minute hour
- Christianize Math
- Mathematize Christianity

25 No-Frills Christmas Crackers

. . . .

1. Which of the letters have a pair of parallel line segments?

   MERRY CHRISTMAS

2. Is the following statement True or False?

   $-25.25°C$ is hotter than $-25.52°C$

3. A group of 25 friends met at a school's reunion. Each person shook hands with the other twenty-four. How many handshakes were exchanged?

4. What is the largest 25-digit number that can be divided by 2 and 5 without any remainder?

5. A square of side 25 cm has four of its corners cut off. What is the perimeter of the new figure?

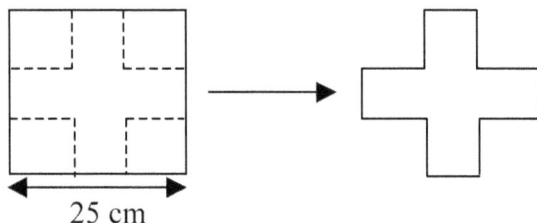

25 cm

6. Tomorrow is Sunday. What day is 25 days after 25 days before yesterday?

7. A carpenter takes 60 minutes to saw a piece of log into six pieces. How long will it take him to saw an identical log into 25 pieces?

8. Weather forecasters: "The temperature fell to *minus 25 degrees*." Mathematicians: "The temperature fell to *negative 25 degrees*." Who are correct?

9. Determine the total number of distinct shortest routes from point A to point B in the following 5 by 5 grid map.

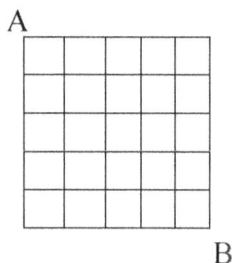

10. The crossed figure is made up of 6 squares. Each side is 5 cm.
    (a) What is the perimeter of the figure?
    (b) What is the area of the figure?

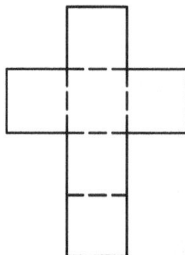

11. Evaluate $\dfrac{1}{1\times2} + \dfrac{1}{2\times3} + \dfrac{1}{3\times4} +\cdots+ \dfrac{1}{24\times25}$.

> Look at a simple case.

12. A rectangular chocolate bar is made up of 25 pieces. Find the smallest number of snaps needed to break the bar into its individual pieces.

> Can I *Google* the answer?

13. What is the value of

$$\frac{1}{\sqrt{1} + \sqrt{2}} + \frac{1}{\sqrt{2} + \sqrt{3}} +\cdots+ \frac{1}{\sqrt{24} + \sqrt{25}}?$$

14. Look up the calendar to find what day of the week December 25, 2010 is. Use this information to find what day of the week December 25, 2015 will be.

| Dec |
| 25 |

15. The following are screen captures from a graphic calculator (T1-84). Match each graph with its corresponding equation.

A. $y = x^{25}$
B. $y = \dfrac{1}{x^{25}}$
C. $y = x^{25} + \dfrac{1}{x^{25}}$

 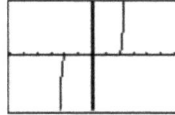 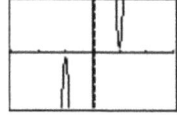

A   B   C

16. Given a 25-piece square chocolate bar, how many snaps are required to break the bar into its individual pieces?

17. In how many ways can a debt of $25 be paid exactly using only $2 bills and $5 bills?

18. On Christmas day Santa Claus leaves you this note:

$$3(\text{Water} - 2)$$

What does it mean?

19. Using only two straight cuts, divide the cross on the right into three pieces and reassemble them to form a rectangle twice as long as it is wide.

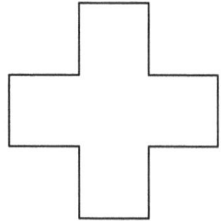

20. Joseph took 2.5 days to make a cross. Mary took 4 days to make a similar cross. Working at these rates, how long will both take to make 25 crosses?

21. Mary folds a piece of paper. How should she draw the figure on the paper so that she can get the figure on the right?

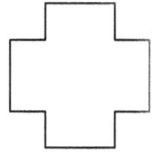

22. A cross is made up of five congruent squares. If $XY = 10$ cm, what is the area of the cross?

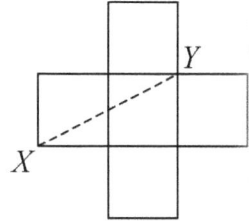

23. Santa's score in a mathematics contest is in the 25th from both the highest and the lowest scores. How many contestants were there?

24. The figure shows an arrangement of 20 points. How many squares can be formed with any 4 of the points as vertices?

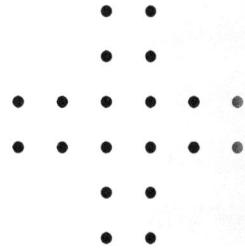

25. The symbolism $\lfloor x \rfloor$ denotes the largest integer not exceeding x. For example, $\lfloor 3 \rfloor = 3$ and $\lfloor 11/2 \rfloor = 5$. What is $\lfloor \sqrt{1} \rfloor + \lfloor \sqrt{2} \rfloor + \lfloor \sqrt{3} \rfloor + \cdots + \lfloor \sqrt{25} \rfloor$

Did you know...

- Christmas reminds believers of Christ's birthday and that he was born some 2000 years ago in the city of Bethlehem.

- Christmas and Easter Christians are nominal believers who fulfill their religious duty every year to appear spiritual.

- Between the seventh century and 1338 the English considered Christmas Day to be the start of the new year, but in 1339 they moved it to 25 March for civil purposes and Easter for religious purposes.

- The town of Santa Claus can be found in southern Indiana, USA. It has a 7-meter-high statue of Santa Claus surrounded by fields of maize.

- The British mathematician and scientist Isaac Newton was born prematured on Christmas Day, 1642.

- The last solar eclipse on 25 December occurred in 1954. The next Christmas Day solar eclipse will not happen until 2307.

- The first Christmas card was printed in the US in 1875 by Louis Prang, a Massachusetts printer.

- The first postage stamp to commemorate Christmas was printed in 1937 in Austria.

- If every day were Christmas Day... you would now be several thousand years old.

The Mathematics of Christmas

For a moment, let our child-like imagination run wild into believing that the mythical, white-bearded fatherly figure, Santa Claus, does exist. And let's look at the myth and the math of Christmas. How can Father Claus make millions of children happy every Christmas? How does he know where children live, and what toys they want? How can he fly in any weather, circle the globe overnight, carry millions of toys and make numerous visits? How many reindeer would Father Claus require?

The United Nations Children's Fund (UNICEF) tells us there are some 2106 million children—about two billion—below 18 years of age in the world. Given that Christmas is no longer celebrated by Christians only, we can assume that Santa will deliver presents to children of all faiths, and not just the 200 million who live in developed countries.

> Be a Santa Claus, today!
> http://www.unicef.org/

Let each household have an average of 2.5 children. Father Claus has to make some $(2,106 \div 2.5) \approx 842$ million stops on Christmas Eve. And suppose these homes are spread equally across the land masses of the planet.

With a radius of 6,400km, the Earth's surface area is about 510,000,000sq km (510 million sq km). Since about 29 percent of the surface of planet Earth is land, this leaves us with a populated area of 150,000,000sq km. Each household therefore occupies an area of 0.178sq km ($= 150,000,000 \div 842,000,000$). Let's assume that each home occupies the same sized plot, so the distance between each household is the square root of the area, which is 0.42km.

Time of Delivery

Every Christmas Eve, Santa has to travel a distance equivalent to the number of visits—842 million—multiplied by the average spacing between households, which turns out to be 356 million km (= 842 million × 0.42.) Fortunately, to cover this distance, Santa has more than 24 hours in which to deliver the presents, if he takes advantage of the different time zones.

Consider the first point on Earth to go through the *International Date Line* at midnight on 24 December. From this moment on, if Santa stays right there, he will have 24 hours to deliver presents to every home along the Date Line. But he can do better than this, by traveling backwards, against the direction of rotation of the Earth. That way he can deliver presents for almost another 24 hours to everywhere else on the planet, making 48 hours in all, which is 2,880 minutes, or 172,800 seconds.

International Date Line is the imaginary line running from north to south through the Pacific Ocean, east and west of which the date differs by one day. The line is 180° from Greenwich in England.

Speed of Sleigh

To get between each of the 842 million homes, Santa has a little over 0.000205 second (= 172,800 ÷ 842,000,000) to do so. To cover the total distance of 356 million km in this time means that his sleigh is moving at an average speed of 2,060km per second. Given that the speed of sound is about 1,200km per hour, Santa is achieving speeds close to 6,180 times the speed of sound.

Weight of Toys and Number of Reindeer

Assuming that each of the 2,106 million children gets nothing more than a toy of average size, weighing, say, 900g, Santa has a load of 1,895 million kg (= 2,106 × 0.9 million) to deliver. On average a reindeer can carry about 135kg, and assume that a flying reindeer can carry ten times more load than a normal reindeer. Then Father Claus would need about (1,895 million ÷ 1,350) = 1,400,000 reindeer to transport all the Christmas presents.

Santa's beloved deer

The Traveling Santa Problem

The traveling Santa has to visit a number of cities (say, N) each only once, while minimizing the total distance traveled. If the number of cities is five, Santa has five ways to choose the first city. Upon visiting the first city, he has four ways left to visit the second city; then three more ways to visit the third city, and so on. So, by the multiplication principle of counting, he has a total of $1 \times 2 \times 3 \times 4 \times 5 = 120$ possibilities to travel through the five cities. With 10 cities, there would be $(1 \times 2 \times 3 \times \cdots \times 9 \times 10) = 3,628,800$ possibilities. With just 25 cities, a computer evaluating a million possibilities per second would take 490 billion years—40 times the age of the universe, to search through them all.

For the 842 million households, Santa would take 10 to the power of 7.15 billion years. (10 to the power of 2 is 10×10, denoted by 10^2; 10 to the power of 5 is 10^5, and so on.)

A *CHRISTmaths* Wish

While the myth of Santa Claus provides us with a magical time merrymaking, the math of Christmas disproves his physical existence. The miracle of Santa, his sleigh and all of his reindeer to deliver the toys to all children of the world every Christmas Eve, may not remain science fiction for eternity, if we speculate what or where some of the present advanced scientific theories or hypotheses (DNA computers, wormholes, time machines, nanotechnology, and so on) may bring us.

The science and mathematics of Santa Claus may turn the myth of Christmas into some reality in the future. Who knows? There could be billions or trillions of Santas out there delivering presents in parallel universes, besides planet Earth.

Merry *CHRISTmaths* to everyone.

Practice

0. Given a radius of 6,400 km, show that the Earth's land surface area is about 150,000,000 km². [Area of sphere = $4\pi r^2$]

1. Verify that Santa's sleigh is moving at an average speed of 7,416,000 km/h.

2. Given that the speed of light is about 300 million meters per second, show that Santa is traveling close to 1/145 the speed of light.

3. (a) If Father Claus were to visit five cities, he could do it in 120 ways. How many possible ways would there be if he were to travel to seven cities?
    (b) If a computer can evaluate a million possibilities per second, verify that with 25 cities it will take about 490 billion years to search through all possible ways.

4. "Jesus Christ was not born on December 25 in the year 1 AD." Comment on the numerical fact or myth.

5. Guesstimation
   (a) How many Christmas cards are sent every year round the globe?
   (b) How many children round the world receive presents every Christmas?

6. Out of the two billion children (persons under 18) in the world, if Santa does not handle the Buddhist, Hindu, Jewish, and Muslim children, that reduces the workload to 15% of the total 378 million according to the Population Reference Bureau. At an average rate of 3.5 children per household, how many homes will that be?

Father Christmas deserves a rest this year. He never gets a sick day!

You can't fool Santa! He knows your every naughty act!

25 Quickies & Trickies

1 How Many Ways

How many ways can 25 books be arranged on a shelf?

2 A Sum of Odd Terms

Without a calculator, or adding the terms, one by one, what is the sum of

$S = 1 + 3 + 5 + \cdots + 25?$

4 The Largest 25-Digit Integer

What is the largest 25-digit number that can be divided by 2 and 3 without any remainder?

3 The 25th Power of 2.5

If $25^{25} \approx 8.881784197 \times 10^{34}$, what is the value of 2.5^{25}?

5 A Christmas Call

"Hi, is this 9001-3267?"

"Yes, who's on the line?"

"Don't you recognize my voice?"

"My mother is your mother's mother-in-law!"

Who is on the phone?

6 A Sum of Alternating Terms

Find the sum
$S = 1 - 2 + 3 - 4 + \cdots + 25.$

7 A Group of 25 Nations

Is it possible to list 25 countries, each beginning with a different letter of the alphabet?

8 Perfect 25

For mathematical reasons, in codes and ciphers it is common to have 25 (which is a perfect square) letters rather than the usual 26 letters of the alphabet. Which letter of the English alphabet is left out and why?

9 The Number of Subsets

How many subsets does a 25-element set have?

10 The Christmas Star

What is the Christmas star commonly known as?

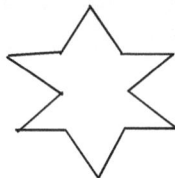

11 A Finite Sum

Find the sum
$$S = \frac{1}{1 \times 2} + \frac{1}{2 \times 3} + \frac{1}{3 \times 4} + \cdots + \frac{1}{24 \times 25}$$

12 The Number of Ways

In how ways can you arrange the letters of the word **CHRISTMASTIME**?

13 A Sum of 25 Powers

Find the sum of
$$S = 2 + 2^1 + 2^2 + 2^3 + \cdots + 2^{25}.$$

14 A Circular Seating Arrangement

There are 25 persons sitting round a big circular table. How many of the $25 \times 24 \times \cdots \times 3 \times 2 \times 1$ arrangements are distinct, i.e., do not have the same neighboring relations?

16 Powers of 2

Express 2^{25} as the sum of two consecutive odd integers, giving your answer in powers of 2.

15 Consecutively Yours

Find 25 consecutive composite numbers consecutive—no primes between them.

17 Another Finite Sum

Find the sum

$$S = \frac{1}{1\times3} + \frac{1}{3\times5} + \frac{1}{5\times7} + \cdots + \frac{1}{23\times25}.$$

18 Areas & Perimeters on a Grid

(a) On a 5-by-5 array of dots, find two triangles with equal areas but unequal perimeters.

(b) How many such pairs of rectangles are there in the 5-by-5 array?

19 Divisibility by 25

Show that for natural number n, $72^{2n+2} - 47^{2n} + 28^{2n-1}$ is divisible by 25.

20 Another Nested Radical

Given that x is a real positive number,

solve $\sqrt{x + \sqrt{x + \sqrt{x\cdots}}} = 25$.

21 **Number of Rectangles**

A board has 25 squares. How many individual rectangles are there in all?

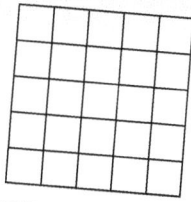

22 **A Sum of Reciprocals**

Find the sum of
$$S = 1 - \frac{1}{2} + \frac{1}{2^2} + \frac{1}{2^3} + \cdots - \frac{1}{25^2}$$

23 **A Nested Radical**

What is the value of
$$\sqrt{25\sqrt{25\sqrt{25\sqrt{25\cdots}}}}\,?$$

24 **Nesting 2's and 5's**

Show that
$$\sqrt{2 + 5\sqrt{2 + 5\sqrt{2 + 5\sqrt{2\cdots}}}} = 5 + \cfrac{2}{5 + \cfrac{2}{5 + \cfrac{2}{5 + \cdots}}}$$

25 **A Sum of Squares**

Find the value of
$$1^2 - 2^2 + 3^2 - 4^2 + \cdots + 25^2.$$

Was Pythagoras a pre-Christian Christian?

A mystical parallelism between Mathematics and Christianity involves looking at some similarities and differences between Pythagoras and Jesus Christ.

Pythagoras	Jesus
1. Son of the god Apollo by a virgin birth to his mother, Pythais.	1. Son of God by a virgin birth to his mother, Mary.
2. Pythagoras's father, Mnesarchus, received the glad tidings from the Delphic oracle.	2. Jesus' earthly father, Joseph, was told of his birth by the angel Gabriel in a dream.
3. Semi-divine; Did miracles, conversed with daemons, and heard the music of the stars.	3. Divine; Did miracles, cast away demons, and heard from the Father directly.
4. Pythagoras conveyed the reality of the spiritual life.	4. Jesus conveyed the reality of the spiritual or eternal life.
5. Pythagoras himself left no written record; his disciples, the Pythagoreans, transmitted his teachings.	5. Jesus himself left no written record; the records were written by his disciples and followers.
6. Pythagoras contemplated on a holy mountain in Crete where he was initiated into the secret rites of Idaean Zeus.	6. Jesus contemplated 40 days in the wilderness with the Father.

7. Pythagoras spread his teachings in the form of parables, called *akousmata* by the Pythagoreans.	7. Jesus spoke in parables, as recorded in the New Testament by his disciples.
8. Pythagoras suffered a terrible death; he was killed in the fields by the common people. (for intellectual snobbery).	8. Jesus was crucified by his own people who refused to receive his gospel (for proclaiming that He Was the Son of God and Has the power to forgive sins).
9. Pythagoras was said to have ascended bodily into the heavens upon his death.	9. Jesus was resurrected after his death to fulfill what the prophet Isaiah predicted in the Old Testament.
10. Pythagoras preached the reincarnation of life.	10. Jesus preached the resurrection of life.
11. Pythagoras believed in the evolution of man.	11. Jesus believed in the creation of man.
12. Pythagoras saw man as a rational animal.	12. Jesus saw man in the image of God (Genesis 1:27).
13. The cosmology of Pythagoras contradicts that of the Book of Genesis.	13. The cosmology of Jesus supports the Book of Genesis.
14. Pythagoras's disciples revered the tetractys.	14. Jesus' disciples use the cross to remind them of their Master's sacrifice.
15. Pythagoras's belief that numbers rule the universe ushered in a new age of scientific thinking away from superstition and myth.	15. Jesus' earthly ministry and the promise of the Holy Spirit ushered in a new age of moral conduct, away from good works and religious rituals.

16. Pythagorean numerologists seek to exploit numerical relationships as a way to understanding the workings of the universe	16. Biblical numerologists seek to exploit numerical symbols as a way to demonstrating the inspiration and deep harmony of the Scriptures.
17. Present-day Pythagorean groups or cults include the Freemasons and the Rosicrusians.	17. Present-day Christian unorthordox groups include the Mormons, Seven-Day Adventists, and Christian Science.
18. Pythagoras's maxims (his golden verses) laid down esoteric practices for his disciples: "Eat not the heart." "Receive not swallows into your house." "Do not carry the images of the gods in rings." "Refrain from wearing wool clothing." "Never touch a white rooster." "Never sit on a quart measure."	18. Jesus' wise sayings revolutionized the current way of thinking among the people then, and unveiled new standards for his followers: "Love your enemies." (Mt 5:44) "A man who looks at a woman lustfully has already committed adultery with her in his heart." (Mt. 5:28) "Whoever who loves his father or mother more than me is not worthy of me." (Mt. 10:37) "If someone strikes you on the right cheek, turn to him the other also." (Mt. 5:39) "Bless those who persecute you." (Rom. 12:14)

19. Pythagoras's dictum: "All is number."	19. Jesus' declaration: "No one comes to the Father except through Me."
20. Belief in Pythagoreanism paves the way to the numerological world.	20. Belief in Christology paves the way to the demonic or gnostic world.
21. The number 10 (the decad) underlied the Pythagorean theology.	21. The number 7 is covertly hidden throughout the entire Bible.
22. The Pythagoreans were superstitious of the number 17.	22. The Christians are superstitious of the number 666.
23. Pythagoras preached that one should not eat animals or beans because they had souls.	23. Jesus preached that human beings, and not animals or things, have souls.
24. Pythagoras ordered his disciples to lead an ascetic life. The Pythagoreans lived simply and poorly, with members sharing their belongings.	24. Jesus taught his disciples to lead a simple life, not to waste their lives on material things by accumulating earthly riches which they cannot bring in the next life.
25. Members of the Pythagoras' Academy recognized each other by using secret signs (e.g., the Pentagram.)	25. Members of the Christian Church recognize each other through the gifts of the Holy Spirit (e.g., speaking in tongues, prophecy, …).

A Formula for Christmas Day

· ·

Pretend to be Magician Santa, and impress your friends by revealing the day of the week on which Christmas Day falls in any year.

On what day of the week does Christmas Day fall in the year 2118? Or in 2025? Mathematicians have come up with an idiots-proof cute formula that can reveal the day of the week on which Christmas Day falls in any year (including leap years).

1. Divide the year you are interested (say, 2118) into its century number C (= 21) and its year number Y (= 18).

2. Divide C by 4, and retain the whole number part, denoted by K. So, K = 21 divided by 4 = 5.25, which is rounded off to give K = 5.

3. Do the same for Y, keeping the whole number, denoted by G. So, in our case G = 18 divided by 4 = 4.5, rounded off to 4.

4. Work out the value of D, using the formula $D = 50 + Y + K + G - (2 \times C)$.

In our case, $D = 50 + 18 + 5 + 4 - 42 = 35$

| Dec |
| 25 |

5. Divide D by 7, and write down the remainder, R. Use the
    following table to give the day of the week on which Christmas
    falls:

R	Day
0	Sunday
1	Monday
2	Tuesday
3	Wednesday
4	Thursday
5	Friday
6	Saturday

How on earth did those folks come up with such a formula?

In our case, D = 35, so D divided by 7 is 5, with no remainder
(R = 0).

This means Christmas Day 2118 will fall on a Sunday.

The challenge here is to figure out how the formula for Christmas Day
came about. Are you ready to read the mathematician's mind?

Practice

1. On what day of the week does Christmas Day fall in the
    following years?
    (i)   2004   (ii) 1986
    (iii) 2025   (iv) 2215

*2. Investigate: What day of the week is Christmas Day most likely
    to fall in a 400-year cycle?

**3. Design your own formula for Christmas Day that spans into the
    4th millennium.

Q&A about Christmas

Why retailers put Christmas decorations on display in September?

Shops display their Xmas goods as early as late September or early October purely due to economic considerations. On average, the holiday season accounts for about 40 percent of annual retail sales volume and almost 65 percent of annual retail profits.

> Nearly one-fourth of all yearly sales come from Christmas purchases!

As a result, shelves used for holiday merchandise cannot be used to display other merchandise. So one needs to be creative during the festive season to avoid having one's products being booted off the shelves!

In the United States, retail sales during December tower visibly over the volume in adjacent months. In some categories (with familiar Yuletide wares), December sales account for a huge share of the year's sales, over a fifth at jewelry stores, about a sixth at departmental and discount stores, and about a seventh at clothing, electronics, sporting goods, hobby, and book stores.

Is "X'mas" a proper abbreviation for Christmas?

Yes, it goes back to Old English—the Greek word "Christ" begins with the letter *chi*, or *x*).

Why is one more likely to die during Christmas than during the rest of the year?

I don't believe in Santa but do believe in the spirit of Santa.

Heart-attack patients have higher mortality rates than those admitted during the rest of the year. During the holiday period patients are more likely to receive second-rate drugs or surgery. Why? The best doctors and surgeons have made sure they get time off, leaving people unfortunate enough to get coronaries for Christmas in the hands of less able staff.

So the next time you plan your surgery, avoid having it during the festive season, when surgeons prefer to be with their family members and friends than with their patients.

When do the twelve days of Christmas begin?

The Twelve Days of Christmas start on December 26 and end on January 6 (Epiphany).

What is the significance of the twelve days of Christmas?

One interpretation for the twelve days of Christmas is tabulated below.

The 12 Days of Christmas

Day	Gifts from True Love	Christian Interpretation
1st	A Partridge in a Pear Tree	*The One true God*
2nd	Two Turtles Doves	*The Old and New Testaments*
3rd	Three French Hens	*Faith, Hope, and Charity*
4th	Four Calling Birds	*The Four Gospels*
5th	Five Golden Rings	*The Books of Moses*
6th	Six Geese a-Laying	*The Six Days of Creation*
7th	Seven Swans a-Swimming	*The Seven Gifts of the Holy Spirit*
8th	Eight Maids a-Milking	*The Eight Beatitudes*
9th	Nine Ladies Dancing	*The Nine Fruits of the Spirit*
10th	Ten Lords a-Leaping	*The Ten Commandments*
11th	Eleven Pipers Piping	*The Eleven Faithful Apostles*
12th	Twelve Drummers Drumming	*The Apostles' Creed*

How many gifts did my true love send to me on the twelve days of Christmas?

On the first day of Christmas, the beloved received one gift, on the second day 2 + 1 = 3 gifts; on the third day, 3 + 2 + 1 = 6 gifts; and so on.

On the first day of Christmas, she had accumulated only one gift; on the second day she received three more gifts for a total of four gifts. On the third day she received six more gifts for a total of ten gifts, and so on. The total number of gifts is tabulated below.

Day	Number of gifts given on that day	Number of gifts on all days combined
1	1	1
2	3	4
3	6	10
4	10	20
5	15	35
6	21	56
7	28	84
8	36	120
9	45	165
10	55	220
11	66	286
12	78	364

So if the gifts are returned at the rate of one gift per day, it will take 364 days to return all the gifts mentioned in the song "The twelve days of Christmas."

Another way to find the total number of gifts my true love sent to me is shown in the table below.

Item	No. of items given	No. of days given	Total no. of each item
Partridge in a pear tree	1	12	12
Turtle doves	2	11	22
French Hens	3	10	30
Calling birds	4	9	36
Golden rings	5	8	40
Geese-a-laying	6	7	42
Swans-a-swimming	7	6	42
Maids-a-milking	8	5	40
Ladies waiting	9	4	36
Lords-a-leaping	10	3	30
Pipers piping	11	2	22
Drummers drumming	12	1	12
		Total	364

What is the Santa Claus rally?

The advice from investors is to "play the Christmas markets" when there are low liquidity and trading volumes of stocks, a lack of sellers, and most trades are buys.

According to seasoned players, chances of one losing money by investing in the stock market on Christmas Eve and running this position through to the first trading day of January are slim. This is because Christmas Eve, and December 27 and 28, which have relatively light trading, are some of the strongest trading days of the year, as far as trading in the United Kingdom is concerned.

According to experts, for example, the UK market has, on average, risen over 75 per cent of the time in the two trading days prior to Christmas and the three trading days following Boxing Day.

So, if you are an investor, you can apparently improve the odds of a happy Christmas by playing the Christmas markets, when sellers are more interested in Christian festivities than in analyzing any bull or bear rallies. It is also when Wall Street's buyers have a greater influence than normal.

Why not draw a *line graph* to depict the trend of the stock market during the twelve days of Christmas?

Clausophobia and the Rest

Fear of math = Mathophobia

Fear of the number 25: [Unknown]

Fear of Christmas: *Xmasphobia*

Fear of Father Christmas: Santaphobia

Fear of toys: Toysphobia

Fear of crosses or the crucifix: Staurophobia

Fear of giving birth to a Christmas baby: *Jesuphobia*

Fear of not receiving a Xmas present: [Unknown]

Fear that Santa actually exists: *Clausophobia*

Fear of Christmas trees: Christougenniatikodentrophobia

Fear of attending a Christian church service: *Massphobia*

Fear of being invited to a Christmas Eve's service: *Xmassphobia*

Fear of being dubbed a Christmas church believer: *Xtianphobia*

Fear of sharing the Good News of Jesus Christ: *Gospelophobia*

Fear of offending non-believers on Christmas Day:

> Santa Claus is an improbable source of improbable wants.

> Fear of reading all those Xmas phobias?

Santa Claus

Christmas gift giving is a cross that we must bear.

Why, Santa couldn't be real

How Satan kidnapped Santa

Most-wished Items 4 Xmas	
1980s	2000s
Toys	Cash Gift cards
Books	Clothing DVDs
Walkmans	iPhones

For math solutions, claim JAMES 1:22.

The Yule-log

Bank's Xmas Offer 2.5% Interest ONLY

25 Things NOT to do on CHRISTMAS

In Germany
We Wish You a Buried Christmas!

CROSS - curses = Blessings

Conversion Rate	
Year	# of new believers per minute
1825	45
1925	130
2000	186
2025	?

* Assuming His Second Coming is after 2025.

Is Christmas a necessity or luxury?

More deaths on Xmas! Reason: 25-hour workers

Mathematical Graphiti

A Xm●s wh●lly d●y
●r
A Xm●s h●ley d●y?

Keep the Christ in Christmas!

⚲ ✝ m @ 5

Jesus was 1999 years old in 1996!

There was no zero year.

The Star of Bethlehem was a comet!

gold, frankincense and myrrh

Xmas Presents

Batteries – toys not included

The One Hundred Days of Christmas

What Would Santa Do? (WWSD)

virgin birth: XY – Y

A CHRISTmaths
— Gift —
Prove that the total number of gifts my true love sent to me is 364.

Jesus is coming back. Look busy!

Father Christmas hanged before 25 Sunday school children

25 Fiends & Foes
I Want 2 Pray 4 r:

Judas   Superman
Lucifer   ...

Santa must die! Christ lives!

From pregnancy to eternity

Dec. 25 ——Jn 3:16——→ ∞

christillion
1,000,...,000
(25 zeros)

Magi ≡ 3 wise men & a baby
Marygi ≡ 3 wise women

81

Number of Zeros in 1 x 2 x 3 x···x 25

A few Christmases ago I e-mailed the following festive cracker to tickle oft-math-averse fellow colleagues, accompanied by an informal solution to this nonroutine question.

> **How many zeros are there at the end of the product?**
> $$1 \times 2 \times 3 \times \cdots \times 23 \times 24 \times 25$$
>
> No calculators allowed.

Thou shalt not let the devil frighten you with the above grade five olympiad math question!

Let's solve the Xmas cracker together. Of course, Monsieur Lucifer won't mind you spending twenty-five odd minutes multiplying all twenty-five numbers, until your mental calculator runs out of steam. No, don't listen to his (or her or its—not sure about the devil's gender, though!) mathematical white lies.

Using some elementary arithmetic, we can painlessly work out the answer—yes, even for those of you who flunked your nursery math.

Question 1: When will multiplying two numbers end in a zero? When the units (or ones) digits are 2 and 5, isn't it?

Question 2: How many numbers ending in a 5 are there in $1 \times 2 \times 3 \times \cdots \times 24 \times 100$? I mean numbers that can be rewritten as a product of 5. A no-brainer again: 5, **10** (2×5), **15** (3×5), **20** (4×5) and **25** (5×5). Note that the two 5's from 25 can be multiplied by any even number (2, 4, 6, …, 24) to give two products, each ending with a zero.

So, there are in all six 5's (four from 5, 10, 15, 20 and two from 25) which can be multiplied by any even numbers from 1 to 25 (12 of them, if you want to verify) to give six intermediate products, all ending in a zero.

Note that we needn't worry about the number of 2's (or evens), because there are enough of them between 1 and 25 to marry the six fives.

Now, since any number multiplied by 10 or multiples of 10 (like 20, 30, 40. ...) will always lend a zero to the final product, so the product of $1 \times 2 \times \cdots \times 24 \times 25$ will contribute 6 zeros by marrying 6 fives and 6 evens. Get it? I prayerfully hope so!

Hence, there are exactly 6 zeros in the product $1 \times 2 \times 3 \times \cdots \times 24 \times 25$: #*#*#*... **000,000**.

Can you excitedly say "*Aha! I get it!*"?

Merry X-W/g 2 All!

YKC—Your Kidding Claus

Postscript: This is *Question 25* taken from a quasi-completed manuscript of 25×25 questions. The author of **CHRISTmaths** is looking for a publisher—he promises to donate all royalties in the form of maths books to dozens of children living in some remote parts of Africa.

Two Less Informal Solutions

How many zeros are there at the end of the product,

$$1 \times 2 \times 3 \times \cdots \times 24 \times 25?$$

Method 1

If $1 \times 2 \times 3 \times \cdots \times 24 \times 25$ is written as a product of factors, it will contain six 5's and more than six 2's among many factors:

$1 \times 2 \times 3 \times (2 \times 2) \times 5 \times (2 \times 3) \times 7 \times (2 \times 2 \times 2) \times (3 \times 3) \times (2 \times 5)$
$\times \cdots \times (3 \times 5) \times \cdots \times 19 \times (2 \times 2 \times 5) \times (3 \times 7) \times \cdots \times (5 \times 5)$

Then part of the product can be written as:

$$(5 \times 2)(5 \times 2)(5 \times 2)(5 \times 2) \ (5 \times 2)(5 \times 2),$$

which can also be represented as $10 \times 10 \times 10 \times 10 \times 10 \times 10$.

Therefore there are six zeros in the product $1 \times 2 \times 3 \times \cdots \times 25$.

Method 2

Another method is to use a mathematical function what mathematicians call the *greatest integer function* (short for the "greatest integer less than or equal to" function) but computer scientists call it simply "floor" and many calculators call it **int**.

We can think of this function as "rounding down." For positive numbers we simply drop the fractional part; for negative numbers, however, rounding down doesn't work that way. Thus, int $(\pi) = 3$, but int $(-\pi) = -4$.

In general, if we want the whole number quotient of a number N divided by D, we would enter int (N/D). For instance, int $(13/2) = 6$; int $(78/10) = 7$.

For any positive integer N, the remainder, R, for N divided by D is $R = N - D * $ int (N/D).

Let's state a useful theorem that quicky allows us to find the terminal zeros of $n!$

Theorem (Legendre)
Let p be a prime. Then the largest power of p that divides $n!$ is $\lfloor n/p \rfloor + \lfloor n/p^2 \rfloor + \lfloor n/p^3 \rfloor + \cdots$, where $\lfloor x \rfloor$ denotes the greatest integer not exceeding x.

$n! = 1 \times 2 \times 3 \times \cdots \times (n-1) \times n$

How many zeros does 25! end in?

This is equivalent to finding the largest power of 10 that divides 25!

It suffices to apply the above theorem to the prime 5, obtaining

$$[25/5] + [25/5^2] = 5 + 1 = 6$$

How many zeros does 169! end in?

$[169/5] + [169/25] + [169/125] = 33 + 6 + 1 = 40$ zeros

How many zeros does 500!/200! end in?

The largest power of 10 that divides 500! is $[500/5] + [500/25] + [500/125] = 124$.

The largest power of 10 dividing 200! is $[200/5] + [200/25] + [200/125] = 49$.

Thus 500!/200! ends in $124 - 49 = 75$ zeros.

Practice

1. Find the number of terminal zeros in each of the following products.
   (a) $30 \times 300 \times 3000$
   (b) $500 \times 60,000 \times 2500$
   (c) $1 \times 2 \times 3 \times 4 \times 5 \times 6 \times 7 \times 8 \times 9$
   (d) $3 \times 6 \times 9 \times 12 \times 15 \times 18 \times 21 \times 24$
   (e) $5 \times 10 \times 15 \times 20 \times 25$

2. How many terminal zeros are there in the product
$$1 \times 2 \times 3 \times \cdots \times 16 \times 17?$$

3. If the product $1 \times 2 \times 3 \times \cdots \times 29 \times 30$ is written as a product of factors $(2, 3, 5, 7, \cdots)$, how many 5's will the product contain?

4. How many zeros are there at the end of the following product?
$$1 \times 2 \times 3 \times \cdots \times 20$$

5. How many zeros are there at the end of the following product?
$$1 \times 2 \times 3 \times \cdots \times 99 \times 100$$

6. How many zeros are there at the end of the product
$$2 \times 4 \times 6 \times 8 \times \cdots \times 100?$$

7. How many zeros are there at the end of the product
$$1000 \times 998 \times 996 \times \cdots \times 6 \times 4 \times 2?$$

8. What is the maximum number of times the number 2 will divide into
$$1 \times 2 \times 3 \times \cdots \times 49 \times 50 \text{ exactly?}$$

9. How many positive whole numbers N are there with the property that
$$1 \times 2 \times 3 \times \cdots \times N$$
does not end in a zero?

10. How many zeros does $\frac{100!}{25!}$ end in?

11. How many zeros are at the end of $1 \times 2 \times 3 \times \cdots \times 999 \times 1000$?

12. Determine the number of zeros at the end of $5^{10}!$

25 Math Things You Can Do on Christmas

25. Write a review on Amazon.com, which would lead even mathophobic parents to purchase a book like *CHRISTmaths*.

24. Be a Wikipedia writer by linking 25 most-visited math sites to the subject on *Christmaths*.

23. Pose some alphametics with a Christmas theme.

$$A\overline{)\,\text{MERRY}}^{\text{XMAS}}$$

$$\begin{array}{r} \text{S A N T A} \\ -\ \text{C L A U S} \\ \hline \text{X M A S} \end{array}$$

22. Explain how many terminal zeros there are in
$$1 \times 2 \times 3 \times \cdots \times 24 \times 25$$
to an average 125-month-old child.

21. Entertain yourself to some number freaking:
How many toys are delivered on Xmas Day around the globe?
How many Christmas cards are sent every year?

20. If a couple desires to give birth to a Christmas baby, work out the best time of the year for them to come together more often.

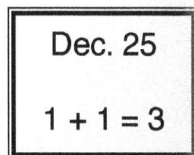

> Dec. 25
>
> 1 + 1 = 3

19. What is the sum of the interior angles of a Christmas tree?

18. **How many new words can you derive from the word CHRISTMAS? You do not have to use all the letters. (Examples: The words *his*, *mast*, and *mist*.)**

17. If you received $1 million from Father Santa and had to spend it all on Christmas, how and where would you spend it?

$$\boxed{\$1,000,000}$$

16. **Prove the nonexistence of Santa Claus.**

Santa is a myth!

We're fiends and foes of Santa!

15. Name 25 words beginning with the letter C that relate directly to the Christmas season? (Examples: *Candle, Compassion, Caring*)

14. **Name the biblical verses that forecast the arrival of Jesus Christ. How fast can you locate them in the Holy Scriptures?**

13. What day of the week would Christmas fall in 2025?

12. **Explain these shortcuts:**

$$
\begin{array}{r}
25 \\
\times\ 25 \\
\hline
625
\end{array}
\qquad
\begin{array}{r}
35 \\
\times\ 35 \\
\hline
1225
\end{array}
$$

$2 \times 3 = 6; 5 \times 5 = 25$
$3 \times 4 = 12, 5 \times 5 = 25$

11. In the song "The Twelve Days of Christmas," how many gifts would one receive if one added up all of the gifts given in every verse from day one to day twelve?

10. **What day of the week would Christmas most likely fall?**

25
★★★

9. What is the probability that one born on Christmas Day would also die on Christmas Day?

8. **Organize a series of Christmas talks during the festive week to popularize mathematics to the public.**

7. Debunk 12 urban myths about Christmas perpetrated by the Church over the centuries to both believers and non-believers.

6. **Decorate a Christmas tree with shiny, silver and gold Möbius strips, glistened with sparkling stars along their surfaces. Visit the Slovenian artist, Teja Krasek's website:**

   http://tejakrasek.tripod.com

> This tree is enough to warm the heart of any romantic mathematician.

5. Sell second-hand toys on eBay® at $XX.25, offering 25% discount for early birds.

4. List 25 similarities between Mathematics and Christmas.

3. Start conceptualizing another best-selling Math book which is at least 25 times more successful than *CHRIST*maths.

2. Set up a rent-a-Santa business.

1. Organize a 12-day group tour for 25 people to Christmas Island.

Think of 25 More
Things You Can Do
on Christmas.

$1 \times 2 \times 3 \times \cdots \times (n-1) \times n$ ends in 25 zeros

First, note that every multiple of 10 from 10 to 100 contributes one zero:

> 10, 20, 30, 40, 50, 60, 70, 80, 90, 100—there are 10 of them.

Plus, every combination of 2 and 5 will each contribute one zero:

> 5, 15, 25, 35, 45, 55, 65, 75, 85, 95—there are 10 of them.

So the total number of zeros thus far is $10 + 10 = 20$.

Note also that 25, 50, 75, and 100 each have two factors, each of which produces another zero when multiplied by any other even number—for example, $25 \times 4 = 100$ and $75 \times 8 = 600$.

In other words, out of the 20 zeros, four were contributed from two 5's:

> 25 (5×5), 50 $(2 \times 5 \times 5)$, 75 $(3 \times 5 \times 5)$ and 100 $(4 \times 5 \times 5)$.

Now there are altogether $(20 + 4) = 24$ of them. This means that $1 \times 2 \times 3 \times \cdots \times 100$ has 24 terminal zeros.

But we need the value of n for which $1 \times 2 \times \cdots \times n$ has 25 zeros. So one more zero from one multiple of 5 greater than 100 (i.e., the number 105) would add up to 25 zeros.

Notice that the next multiple of 5 (i.e., 110) also contributes one zero, thus bringing the total to 26. Therefore, from $n = 105$ to $n = 109$, the number of terminal zeros remains at 25, as $n = 106, 107, 108,$ and 109 do not contribute any extra zeros.

Hence the values of n for which $1 \times 2 \times 3 \times \cdots \times (n-1) \times n$ ends in 25 zeros are 105, 106, 107, 108, and 109.

Practice

1. (a) There are five numbers n such that $1 \times 2 \times 3 \times \cdots \times n$ has exactly 22 zeros at the end. What are they?
    (b) Hence, can you also name the numbers n for which $1 \times 2 \times 3 \times \cdots \times n$ ends with 23 zeros?

2. For what values of n does $1 \times 2 \times 3 \times \cdots \times (n-1) \times n$ terminate in 37 zeros?

3. For what values of n does $n!$ end in 26 zeros?

> $n!$ is read as "n factorial" and is defined as:
> $n! = 1 \times 2 \times 3 \times \cdots \times (n-1) \times n$

4. Find all positive integers n such that $n!$ ends in exactly 40 zeros.

5. Is it possible for $n!$ to end in precisely 35 zeros?

6. Can $n!$ end in exactly 247 or 248 zeros? Explain.

7. Suppose that $n!$ ends in exactly M_n zeros. Show that M_n is approximately $n/4$ for large values of n.

8. What is the smallest positive integer n such that $n!$ ends in at least 2003 zeros? [SMO 2003 (Junior), Q. 28]

9. The number 29! ends in a string of 0's. Let N be the integer that remains when all those 0's are deleted. Find the largest integer k such that 12^k is a divisor of N. [SMO, 2001 (Senior), Q. 29]

10. Find the largest positive integer n such that $n!$ ends with exactly 100 zeros.

Taking up the Cross

· ·

"If anyone would come after me, he must deny himself and take up his cross and follow me. For whoever wants to save his life will lose it, but whoever loses his life for me and for the gospel will save it." Jesus Christ (Mark 9:34-35)

The cross is more than a religious icon, which had creatively been used to pose many interesting puzzles, some of which are devilishly harder than those sterile drill-and-kill questions that bombard millions of school children every day.

A divine symbol

Don't get too crossed if you can't solve a number of these cross challenges, because even number puzzlists and recreational mathematicians failed to come up with an elegant solution to a number of these *toughies* for a long time.

A satanic symbol

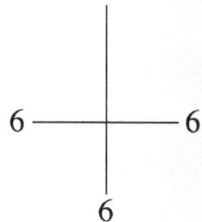

Like matchstick puzzles, these cross problems would serve as a good recreational pastime for convicts (or to those under house arrest) who long to keep their brains active during their long

6 —————— 6

6

solitary confinement. *Who knows?* For some, attempting to solve these problems may eventually lead them to think laterally on the true meaning of the Living Cross—its symbolism and significance to millions of believers who have put their faith in the power of the Cross.

Here are a dozen of cross-like questions to shake your brain. Or, at the least, they could serve as a deterrent to arrest any early sign of senile dementia.

1. Squares in a Cross

A solid Greek cross can be formed by putting together five cubes, or from a number of squares. How many squares are there?

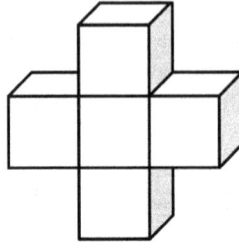

2. Tessellating & Dissecting Crosses

(a) Show how Greek crosses can form a tessellation.

(b) How can an infinite number of dissections from a cross tessellation produce a square?

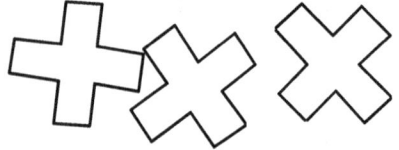

3. A Matchstick Puzzle

The cross on the right is made of 19 matches. Move 7 of them to make a pattern consisting of four squares.

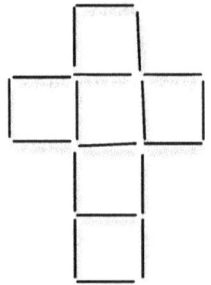

4. Greek Cross Cut

Divide the figure into two identical parts in such a way that when the pieces are rearranged, they form a perfect Greek cross (⊕).

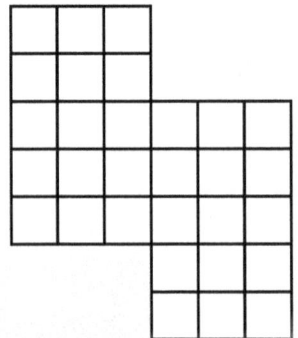

5. Pattern of Crosses

Study the following patterns.

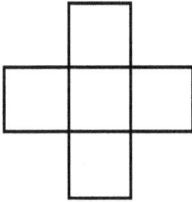
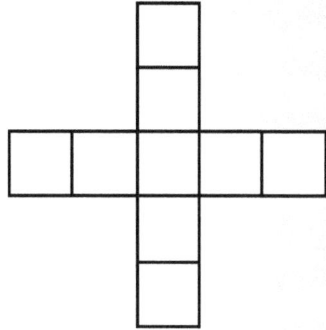

1   2   3

How many squares will be needed to form pattern 25?

6. Rectangle into Cross

Using the seven pieces of the shape below, rearrange them to form a cross.

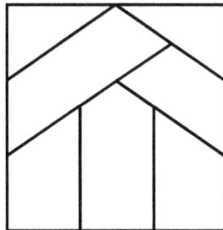

7. Cross into Rectangle

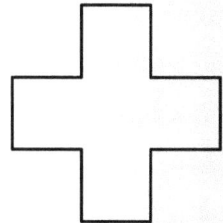

Using only two straight cuts, divide the cross on the right into three pieces and reassemble them to form a rectangle twice as long as it is wide.

8. Five-piece Square into Cross

Cut a square into five pieces and rearrange them to form a Greek cross, as shown below.

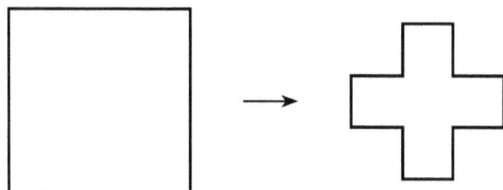

9. Four-piece Square into Cross

Cut a square into four pieces and rearrange them to form a Greek cross, as shown below.

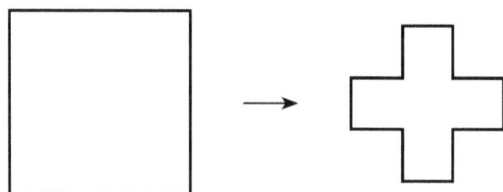

10. Greek Cross into Square

Dissect this Greek cross into nine pieces that can fit together to form either five small squares or one large one.

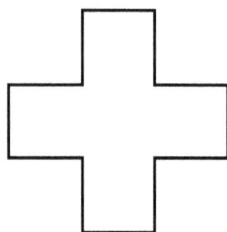

11. Greek Cross into Square

How would you cut the cross into four **similar** pieces and then rearrange them to make a square?

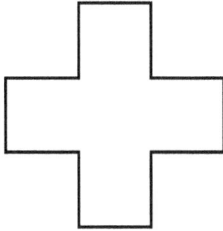

Hint: When the square is assembled, the swastika appears at the centre.

12. Greek Cross into Hollow Square

The Greek cross below has a square-shaped hole in the center.
(a) Rearrange the pieces to make a square that has a hollow cross inside.
(b) Rearrange the pieces so that the result is a square that is rather smaller than the previous "hollow" one.

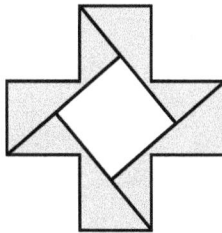

13. Cross into Squares

The cross-shaped figure is made up of five identical squares.
(a) How would you draw four straight lines to cut the figure into pieces that can be rearranged to form a square?
(b) Use the straight lines in a different way to form a square.

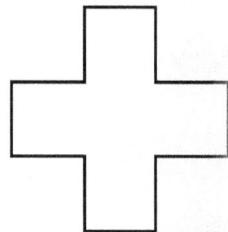

14. The Rolling Disc

Each side of the cross is 10 cm long. A small circular disc of radius 1 cm is placed at one corner, as shown. If the disc rolls along the sides of the figure and returns to the starting position, find the distance traveled by the center of the disc.

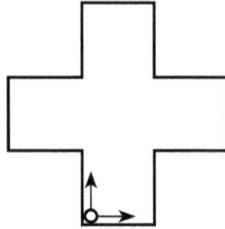

15. The Cross and the Crescent

Reassemble the seven pieces of the crescent to make the Greek cross.

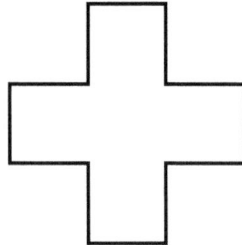

Numerologically
Speaking

JESUS = 888
CROSS = 777
Antichrist = 666

How did they
derive these
numbers?

Mathematicians
Christened

. . . .

The Shakespeare of Mathematics   Leonhard Euler (1707–1783)
The Mathematical Cyclop

The Mathematician Vagabond   Paul Erdös
The Mathematician Monk   (1913–1996)

Paul Erdös

The Mathematician Broadcaster   Marin Mersenne (1588–1648)

The Infidel Mathematician   Edmund Halley (1656–1742)

The Mathematician Diplomat   Gottfried Wilhelm Leibniz
(1646–1716)

The Demigod   Isaac Newton (1642–1727)
The Virgin Mathematician
An Infidel Mathematician

The Mathematician Beaster   Michael Stifel (1487–1567)
John Napier (1550–1617)

God's Mathematical Messenger   Georg Cantor (1845–1918)
and Ambassador

The Jealous Mathematician   Leopold Kronecker (1823–1891)

The Naked Mathematician   Archimedes (287–212 BC)

The Mathematician Adulterer   Bertrand Russell (1872–1970)
The Promiscuous Mathematician
The Mathematician Agnostic

The Mathematician Lover	Évariste Galois (1811–1832)

Évariste Galois

The Prince of Mathematicians *The Mathematical Mozart*	Carl Friedrich Gauss (1777–1855)
The Gambler's Consultant *The Amateur Mathematician-Tutor* The Impoverished Tutor	Abraham de Moivre (1667–1754)
The Devout Geometrician	Blaise Pascal (1623–1662)
The Mathematician Theologian	Bishop George Berkeley (1685–1753)
The Mathematical Mystic	Johannes Kepler (1571–1630)
The Mathematician Womanizer	John von Neumann (1903–1957)
The Homosexual Mathematician	Alan M. Turing (1912–1954)
The Formula Man	Srinivasa Ramanujan (1888–1920)
The Starving Mathematician	Niels Henrik Abel (1802–1829)
The First Mathematics Educator	David Eugene Smith (1860–1944)
The Spiritual Euclid	St. Thomas Aquinas (1225–1274)
The Paranoiac Logician	Kurt Gödel (1906–1978)
The Atheist Mathematician *The Newton of France*	Simon de Laplace (1749–1827)
The Mathematician Paradoxer	Augustus De Morgan (1806–1871)
The Renaissance Mathematician	Leornardo da Vinci (1452–1519)

The Artist Mathematician	M. C. Escher (1898–1972)
The Mathematician Astrologer (Gambler)	Girolamo Cardano (1501–1576)
The Good Samaritan Mathematician	Joseph Louis Lagrange (1736–1813)
The Mathematician Preacher	Thomas Bayes (1702–1761)

Thomas Bayes

The Mathematician Stammerer	Niccolo Fontana (Tartaglia) (1500–1557)
The Amateur Mathematician	Pierre de Fermat (1601–1665)
The Somnambulist Mathematician	Maria Gaetana Agnesi (1718–1799)
The Prophetess of Applied Statistics	Florence Nightingale (1820–1910)
The 23-project Mathematician *The Master Problem Poser*	David Hilbert (1862–1943)

David Hilbert

The "Angelic Doctor"	Peter Ouspensky (1878–1947)
The Gullible Mathematician *The Mathematical "Brooklyn Bridge"*	Michael Chasles (1793–1880)
The Forgetful Mathematician	Lewis Carroll (1832–1898)
The Mathematician Stutterer	Charles L. Dodgson (1832–1898)

The Mathematician Philosopher	René Descartes (1596–1650)

René Descartes

The Mathematical City-Planner *The Fictitious Mathematician* The Cyrano of Mathematicians A "Polycephalic Mathematician"	Nicolas Bourbaki—a collective pseudonym under a (mostly) French group of mathematicians
The Mathematician Terrorist (*aka* the Unabomber)	Theodore Kaczynski (born 1942)
The Mathematician Puzzler	Raymond Smullyan (born 1919)
The Mathematics Expositor & Popularizer	Martin Gardner (1914–2010)
The Mathematical Journalist The Carl Sagan of Mathematics	Keith Devlin (born 1947)
Dean of Recreational Mathematics *The Father of Mathematical Wit*	Charles W. Trigg (1898–1989)
The Mathematical Magpie	John Conway (born 1937)
The "Born Geometer"	Gaspard Monge (1746–1818)
The Schizophrenic Mathematician	John Nash (born 1928)
The Hapless Mathematician	Pierre François André Méchain (1744–1804)
The Modern Calculator	Shakuntala Devi (1939–2013)
The Princess of Parallelograms *The Addicted Mathematician*	Ada Byron Lovelace (1815–1912)
The Mathematician Martyr	Hypatia (370–415 A.D.)
The Mathematician Occultist	Giordano Bruno (1548–1600)
The Grandfather of S-curves	Pierre François Verhulst (1804–1849)

The Fictional Numerologist *The Fictitious Polymath*	Dr. Irving Joshua Matrix (born 1908)
The Poet Mathematician	Omar Khayyam (c. 1050–1123)
The Master Puzzlist	Sam Loyd (1841–1911)
The Anti-Semitic Mathematician	Ludwig Bieberbach (1886–1982)
The Architect Mathematician	Gérard Desargues (1593–1662)
The Eulogist of Mathematics	Roger Bacon (c. 1214–c. 1294)
The Japanese Newton *The "Sacred Mathematician"*	Seki Kowa (1642–1708)
Hitler's Blacklisted Mathematician	Jakow Trachtenberg (1888–1953)
The Mathematics Detective	Sir Arthur Conan Doyle (*alias* Sherlock Holmes) (1859–1930)
The Copernicus of Geometry	Nikolai Lobachevski (1793–1856)
The Maker of Mathematicians	Plato (427–347 BC)
The Machine Man	Charles Babbage (1791–1871)
The Math Counselor	Sheila Tobias (born 1935)
The Multicultural Mathematician	Claudia Zaslavsky (1917–2006)
The Father of Factor Analysis	C. E. Spearman (1863–1945)
The Most Paradoxist Mathematician	Florentin Smarandache (born 1954)
The Math Calendar Teacher	Theoni Pappas (born 1944)
The Non-innumerate Mathematician	John Allen Paulos (born 1945)
The Mathematician Songwriter	Tom Lehrer (born 1928)
The Mathemagician	Benjamin Arthur (born 1961)

Number of Digits in 25^{25}

Earlier on, we learned to determine the number of zeros 25! ends with, and also what values of $n!$ end with 25 zeros. In this section, let's learn to estimate the number of digits in a number expressed in the form of a power. We will also see how to find what number n raised to a given power has a certain number of digits.

Worked Example 1

> Without using a calculator or computer, estimate the value of 25^{25} to a nearest power of 10.

Solution

$$25^{25} = \left(\frac{100}{4}\right)^{25}$$

$$= \left(\frac{10^2}{2^2}\right)^{25}$$

$$= \frac{10^{50}}{2^{50}}$$

$$= \frac{10^{50}}{(2^{10})^5}$$

$$< \frac{10^{50}}{(10^3)^5} \text{ since } 2^{10} > 10^3$$

$$= \frac{10^{50}}{10^{15}}$$

$$= 10^{35}$$

If $a > b$, then $\dfrac{1}{b} < \dfrac{1}{a}$.

So, $25^{25} < 10^{35}$, which is a whole number with 36 digits.

Hence, 25^{25} is roughly a 36-digit number.

Alternatively,

$$25^{25} = (5^2)^{25}$$
$$= 5^{50}$$
$$= (5^{10})^5$$
$$\approx (10^7)^5$$
$$= 10^{35}$$

Thus, 25^{25} has roughly 35 digits.

By calculator,

$$25^{25} \approx 8.881784197 \times 10^{34}$$
$$\approx 10 \times 10^{34}$$
$$= 10^{35}\text{—a whole number with 36 digits}$$

In general, 10^{n-1} is an n-digit number.

Since 10^n (1 followed by n zeros) is the smallest positive integer with $n + 1$ digits, every integer in the interval $(10^{n-1}, 10^n)$ has n digits.

Since $\log 10^{n-1} = n - 1$ and $\log 10^n = n$, an n-digit positive integer has a logarithm in the interval $[n - 1, n)$.

Hence, the logarithm of an n-digit positive integer lies in the interval $[n - 1, n)$.

Let's apply the above result to determine the number of digits in the number 25^{25}.

We can find the number of digits 25^{25} has by taking its common logarithm.

$$\log 25^{25} = 25 \log 25 = 34.9485\ldots$$
$$\text{Thus } 25^{25} \text{ has 35 digits.}$$

Alternatively,
let $25^{25} = 10^n$
$n = 25 \log 25 = 34.94\ldots$
So, $25^{25} = 10^{34.94\ldots}$
Since $10^{34.94\ldots}$ lies the interval $[34, 35)$, 25^{25} is a 35-digit number.

If n is a positive integer such that $\log n \approx 73.1$, how many digits does n have?

Solution

Since 73.1 is in the interval [73, 74), we can conclude that n is a 74-digit number.

Note that we always round up the logarithm of a number to determine the number of digits. In the above case, $\log n \approx 73.1$ is rounded up to show that n has 74 digits.

Practice

1. Without using a calculator or computer, estimate the value of 2^{83} to the nearest power of 10.

2. Without a calculator or computer, which is larger:
$$2^{125} \text{ or } 32 \times 10^{36}?$$

3. Find a whole number n such that n^{24} has 25 digits.

4. What is the smallest integer n for which $25^n > 10^{100}$?

5. Find an integer n such that n^{25} has 2525 digits.

6. Given that $5^{10} = 9765625$,
$$5^{20} = 95367431640625,$$
$$5^{40} = \underbrace{909...625}_{\text{28-digit number}}.$$

Estimate the number of digits in 5^{55}, 5^{555}, and $5^{5^{55}}$.

Christmas Tangrams

Tangram is a 1000-year-old Chinese puzzle that consists of seven pieces arranged in a square: five triangles, one parallelogram, and one square.

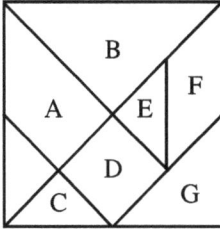

Did you know...

The French emperor, Napoleon Bonaparte was a lover of Tangram!

With these seven pieces, puzzle lovers have creatively formed hundreds of shapes from animals and plants to geometric shapes to items of art.

In China the puzzle is called the "Seven-Board of Cunning" or the "Board of Wisdom," and in Europe the "Chinese Puzzle." The French christened it *Casse-tête*, which literally means "head-breaker."

In a spirit of Christmas, recreational and amateur mathematicians derive much enjoyment forming festive shapes from the seven-piece tangram.

candles   baptismal fonts

Is the number of tangrams infinite?

As you may have guessed, the number of tangrams is infinite!

Strictly speaking, each piece can touch the rest of the tangram at an unlimited number of points. However, the different outlines one would get by putting a given piece in the different positions would not vary much from one another.

Let's restrict ourselves to using the tangram pieces to forming the letters of the alphabet and Arabic numerals, and some geometric shapes.

The Number of Convex Tangrams

In 1942, two Chinese mathematicians, Fu Tsiang Wang and Chuan-Chih Hsiun, asked, and answered, the question, "How many convex tangrams are there?"

A *convex figure* is one that does not have any gaps in its outline. For angular figures (like tangram outlines) this means that all the angles are less than 180°—the corners all stick out instead of in.

Imagine a piece of string or an elastic band being pulled tight around the figure. If this causes the string to make contact with the figure all the way round the edge, then the figure is convex; but if there are gaps between the string and the edge of the figure, then the figure is not convex, but concave, where these gaps occur.

The number of convex tangrams is only a mere thirteen:

Triangles	1
Four-sided figures	6
Five-sided figures	3
Six-sided figures	3

For a proof why there are only 13 convex tangrams, see the paper published in the *American Mathematical Monthly*, Vol. 49 (1942), p. 596.

Practice

1. The Fractional Parts of a Tangram

If the area of the entire square is 1 cm^2, what is the area of each of the seven pieces?

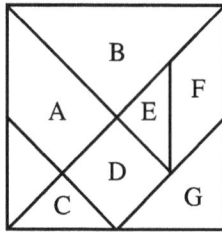

2. The Ten Tangram Numerals

Use the seven pieces of Tangram to form the numerals 1 to 10.

3. The King of kings

Use the seven pieces of Tangram to form these two words.

4. Tangram Paradox

Both silhouettes were made by arranging all seven tangram pieces, but the one on the right seems to have an extra piece. Explain this anomaly.

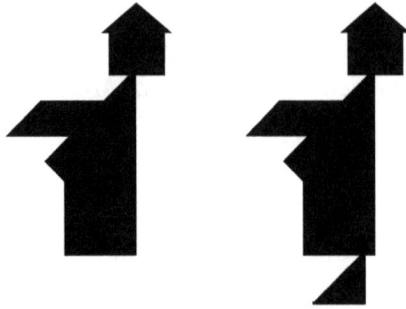

5. Geometric Shapes

Use all seven pieces of the Tangram to form these geometric figures.

(a)

(b)

(c)

6. A Square with a Hole

How would you make the square with the missing square?

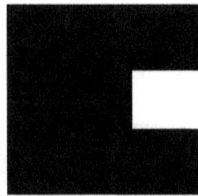

7. The Missing Triangle

Form these two squares, each with a triangular hole.

(a)

(b)

8. The Missing Parallelogram

Form a square with a missing parallelogram.

9. Two Missing Triangles

There are at least 60 different ways to make a square with two triangles missing. Two of them are shown below. Can you form them?

(a)

(b)

10. The Missing Square

How would you make the triangle with the missing square?

CHRISTMAS
By Numbers

Americans send about 2 billion Christmas cards every year, and the average adult posts about 20 cards—the equivalent of 200,000 trees.

More than 25 million kids visit Santa in malls in the U.K. each year.

Every year, 1.76 billion candy canes are made.

Fake Christmas trees have outsold real ones every year since 1991.

Assuming Rudolph's in front, there are $8 \times 7 \times \cdots \times 2 \times 1 = 40,320$ ways to arrange the eight other reindeer.

Odds that a battery was bought during the Christmas season: 40 percent.

Worldwide, Christmas has been celebrated on 135 different days of the year.

The average shopping-center Santa weighs 218 pounds and has a 43-inch waist.

Who still believes in Santa?
Studies say more than four-year-olds do than any other age groups.

Charles Dickens wrote *A Christmas Carol* in six weeks.

Christianity has over a billion followers, followed by Islam with half this number.

It was only after 440 AD that 25 December was celebrated as the birth date of Jesus Christ.

Christmas excess:

> 4 kg CO_2 equivalent per adult
> 280 kg CO_2 equivalent per adult UK average
> 1,500 kg CO_2 equivalent per adult high carbon scenario

The fairy lights burn through about 45 kilowatt-hours.

The average mother spends 13 whole days preparing for Christmas. And after all that, the average Christmas dinner lasts little more than two hours.

Old 1969 calendars could be recycled to be used again in 2025 and 2081.

Keep your 2010 calendar for 2066.

Five and a half million Christmas trees are bought each year.

Compare a 25 cents meal from Cambodia and a $50 buffet from Singapore—a question of inequality.

How many countries in the world have Christmas as a public holiday?

What Day Is Christmas 2025?

In 2010 Christmas Day was on a Saturday. When will Christmas Day next fall on a Saturday? Or, when Christmas Day last fell on a Saturday? Watch out for leap years!

What day was it when you and your parents were born? What day will it be on in the year 2025? In 2525?

Is Christmas Day more likely to fall on a certain day of the week than on other days? In other words, does Christmas Day follow a certain pattern over a period of, say, 400 years?

Let's look at some worked examples to familiarize yourself with the calendar, before you are equipped with the problem-solving strategies to answer some of the above questions.

Worked Example 1

If today is Friday, December 25, and next year is a leap year, what day of the week will it be one year from today?

Solution

First, we need to know the number of days between today and one year from today.

Today's date, Dec. 25, is unimportant except that it shows we are later in the year than the end of December and, therefore, the next year (starting today) will contain 366 days—this would not be so if today were, say, January 25.

Now, $366 = 7 \times 52 + 2$

In modulo arithmetic,
$366 \equiv 2 \pmod 7$

Two days past Friday is Sunday.

Hence one year from today will be a Sunday.

Worked Example 2

If today is Friday, January 25, and next year is a non-leap year, (a) how many days will the next year (starting today) contain, (b) what day of the week will it be one year from today?

Solution

(a) A non-leap year has 365 days, so the following year will have 365 days.

(b) $365 = 7 \times 52 + 1$
     One day past Friday is Saturday.

Worked Example 3

January 1, 1995 was a Sunday. What day was Christmas Day that same year?

1995 is a non-leap year, so February had 28 days.

Month & Date	Number of days
Jan 1 – Jan 31	30
February	28
March	31
April	30
May	31
June	30
July	31
August	31
September	30
October	31
November	30
Dec 1 – Dec 25	25
Total	358

$358 = 7 \times 51 + 1$

One day past Sunday is Monday.

So, Christmas Day in 1995 was a Monday.

Are you ready to solve some calendrical questions?

Practice

1. It is Monday. What day will it be in 25 days' time?

2. January 1 in 1925 was a Thursday.
    (a) What day of the week was December 25 that year?
    (b) What day will December 25 be the following year?
    (c) What day was Christmas Day the year before?

3. What day was Christmas Day 25 years ago?

4. What day will Christmas Day be in 25 years from now?

5. Christmas Day in 2010 falls on a Saturday. When will Christmas Day next fall on a Saturday?

A 20th Century Formula for Christmas Day

Over time, mathematicians have come up with a number of formulas to determine the day of the week Christmas Day will fall in a given year. One such formula is exemplified below.

Worked Example 4

> **Christmas Day in the leap year 19ab (ab is a multiple of 4) is $ab + \dfrac{ab}{4}$ days on from Christmas Day in 1900 (which was on a Tuesday).**

Proof

Let $ab = 4n$

Then $ab + \dfrac{ab}{4} = 5n = 12n \pmod 7 = 4n \times 3 = 3ab$

Moreover, since $3ab + 7ab = 10ab$, we have

$$ab + \frac{ab}{4} = ab \times 10 \pmod 7$$

Applying the above formula to determine the day Christmas fell in 1996, we have

Christmas Day 1996 is 9600 (mod 7) days on from Christmas Day in 1900.

$$9600 = 1 \pmod 7$$

So Christmas Day 1996 is 1 day on from Christmas Day 1900.

But Christmas Day 1900 is on the same day as January 1, 1901, which is 1 day on from January 1, 1900.

January 1, 1900 is known to be on a Monday. So Christmas Day 1900 was a Tuesday, and Christmas Day 1996 was on a Wednesday.

Thus, to find the day for Christmas Day in the leap year $19ab$ is to write down $ab \times 10$, i.e., the number

H	T	U
a	b	0

find the remainder when it is divided by 7, and then count that number of days on from Tuesday (Christmas day in 1900).

What day of the week will be Christmas Day in 2044?

$2044 \rightarrow 1440 \pmod 7$
$1440 = 5 \pmod 7$

5 days from Tuesday is Sunday.

So Christmas Day in 2044 will fall on a Sunday.

Or

$2044 \longrightarrow 20440 = 19$ thousands $+ 1440$

25 Dec. 2044

Sounds like a shortcut?

$25\ 2\ 20440 \div 7 = 3602920 \times 7 + 0$
Remainder $= 0 \rightarrow$ Sunday
Hence, Dec. 25, 2044 is a Sunday.

In general, to find Christmas Day in the leap year $19ab$, calculate $19ab0$ (mod 7) and read off the remainder from the following table.

0	Sunday
1	Monday
2	Tuesday
3	Wednesday
4	Thursday
5	Friday
6	Saturday

For non-leap years

When was Christmas Day in 1998?

To find Christmas Day in 1998, work from Christmas Day in the previous leap year 1996.

Since Christmas Day would have advanced by 1 day in 1997 and by 1 day in 1998, we would need to add 2 to the result for 1996 calculated from 19960 (mod 7).

Since 19962 (mod 7) = 5, Christmas Day 1998 was on a Friday. (Christmas Day in 1996 was on a Wednesday.)

When was Christmas Day in 1995?

To find Christmas Day 1995, we proceed as follows:

1995 is 3 years on from the previous leap year 1992.

Now, 19923 (mod 7) = 1

So Christmas Day 1995 was on a Monday.

To find Christmas Day for a non-leap year is therefore to write down the year of the previous leap year and append on the right as units digit the number of years on from the previous leap year. Find the remainder when this five-digit number is divided by 7, and read off the result from the table of days.

0	Sunday
1	Monday
2	Tuesday
3	Wednesday
4	Thursday
5	Friday
6	Saturday

When is Christmas Day in 2025?

25 2 20241 ÷ 7 = 4 (mod 7)
The remainder is 4, indicating Thursday.

Worked Example 5

On what day of the week will Christmas Day fall in 2025?

Notice that the days of the week from Monday to Sunday repeat themselves every 7 days, or they have a period of 7 days. For example, if today is Wednesday, then 7 days from today will again be a Wednesday, 8 days from today will be a Thursday, and so on.

Determining the day of the week of a given date is an application of what mathematicians called modular arithmetic with modulus 7.

To find the day of the week of a given date, we need to know

(1) the day of the week of any specific date,
(2) the number of days between this date and the date required.

0	Sunday
1	Monday
2	Tuesday
3	Wednesday
4	Thursday
5	Friday
6	Saturday

Let's label the days of the week using the integers 0 to 6 inclusive, as tabulated on the right.

From the 2008 calendar, Christmas Day falls on a Thursday.

From 25 December 2008 to 25 December 2025, there are

$$2025 - 2008 = 17 \text{ full years.}$$

Of these, the leap years are:

$$2012, 2016, 2020, 2024$$

Thus, there are $1 + \frac{1}{4} \times (2024 - 2012) = 4$ leap years.

From 25 December 2008 to 25 December 2025, there are

$$17 \times 365 + 4 = 6209 \text{ days.}$$

Now, $6209 = 7 \times 887 + 0$

The remainder is 0, so Christmas Day in 2025 is again a Thursday.

(a) **What day of the week will Christmas 2025 fall?**

(b) **What will be the next two years when a 2025 calendar can again be reused? In other words, when will be the next two times when Christmas Day will fall on the same day of the week as Christmas 2025?**

(c) **Which previous old calendar can be recycled for a 2025 calendar?**

Solution

(a) Thursday [from Example 5]

(b)

Year	Day of the week on Christmas Day
2025	**Thursday**
2026	Friday
2027	Saturday
2028 (leap)	Monday
2029	Tuesday
2030	Wednesday
2031	**Thursday**
2032 (leap)	Saturday
2033	Sunday
2034	Monday
2035	Tuesday
2036 (leap)	*Thursday*
2037	Friday
2038	Saturday
2039	Sunday
2040 (leap)	Tuesday
2041	Wednesday
2042	**Thursday**
2043	Friday

From the table, the next two Christmases that will share the same day of the week as Christmas 2025 will be in 2031 and 2042.

(c)

2025	Thursday
2024 (leap)	Wednesday
2023	Monday
2022	Sunday
2021	Saturday
2020 (leap)	Friday
2019	Wednesday
2018	Tuesday
2017	Monday
2016 (leap)	Sunday
2015	Friday
2014	**Thursday**
2013	Wednesday

From the table, an old 2014 calendar can be re-used for a 2025 calendar.

Practice

1. Use the 20th century formula for Christmas Day discussed earlier to determine the day of the week Christmas Day fell in those years.
   (a) 1945   (b) 1969   (c) 1999

2. **The Old Calendar**

   In 2008 Grandpa Santa reminded his goddaughter Christ that old calendars could be reused—day by day, a few years later. When would the next year for which a 2008 calendar could be reused?

3. The year 2010 will soon be over. When will be the next year when you can reuse the 2010 calendar?

4. Show that a 2004 calendar may be recycled for the years 2022 and 2042.

5. Modify the 20th century formula for Christmas Day to a 21st century formula to determine the day of the week Christmas Day will fall in a given year.

The Mathematical Fathers

The Father of Deductive Reasoning	Thales (640–546 B.C.) Greek.

Thales

The Father of Numerology	Pythagoras (c. 500 B.C.). Greek.
The Father of Geometry	Euclid (c. 300 B.C.). Greek.
The Father of Logic	Aristotle (384–322 B.C.). Greek.
The Father of Number Theory	Diophantus (c. 250 A.D.). Greek.
The Father of Analytic Geometry/ Modern Philosophy	René Descartes (1596–1650). French.
The Father of Algebra	Al–Khowarizmi (c. 830 A.D.). Arab.
The Father of Infinity	Georg Cantor (1845–1918). German.

Georg Cantor

The Co-Fathers of Calculus	Isaac Newton (1642–1727). English. Gottfried Wilhelm Von Leibniz (1646-1716). German.
The Father of Modern Symbolic Logic	George Boole (1815–1864). English.

The Father of Game Theory	John Von Neumann (1903–1957). Hungarian–American.
The Father of Relativity	Albert Einstein (1879–1955). German–American.
The Father of Platonism	Plato (c. 427–347 B.C.) Greek.

Plato

The Father of Formalism	David Hilbert (1862–1943). German.
The Fathers of Logicism	Bertrand Russell (1872–1970). Alfred North Whitehead (1861-1947).
The Father of Intuitionism	Luitzen Brouwer (1881–1966) Dutch.
The Father of Group Theory	Evariste Galois (1811–1832). French.
The Grandfather of Topology	Leonhard Euler (1707–1783). Swiss.
The Co-Fathers of Probability	Blaise Pascal (1601–1662). French. Pierre de Fermat (1601-1665). French.
The Father of Projective Geometry	Gerard Desargues (1591–1661). French.
The Father of Modern Computing	Charles Babbage (1791–1871). English.
The Father of Descriptive Geometry	Gaspard Monge (1746–1818). French.

The Grandfather of Fuzzy Logic	Bertrand Russell (1872–1970). English.
The Father of Nonstandard Analysis	Abraham Robinson (1918–1974). German.
The Father of Computing Software	Alan Turing (1912–1954). English.
The Father of Quantum Theory	Max Planck (1858–1947). German.

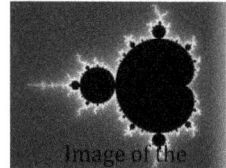

Mandelbrot set

The Father of Catastrophe Theory	René Thom (1923–2002). French.
The Father of Fractals	Benoit Mandelbrot (1924–2010). French.
The Father of Ethnomathematics	Ubiratan D'Ambrosio (born 1932). Brazilian.
The Father of Conjectures and Refutations	Karl Popper (1902–1994). English.
The Father of Modern Statistics	Ronald Fisher (1890–1962). English.
The Father of Lateral Thinking	Edward de Bono (born 1933)
The Father of Linear Programming	George B. Dantzig (1914–2005). American.
The Fathers of Non-Euclidean Geometries	Bernhard Riemann (1826–1866). German. Nikolai Lobachevski (1793-1856). Russian. János Bolyai (1802–1860). Hungarian.
The Father of Modern Problem Solving	George Pólya (1887–1985). Hungarian.

The Answer
Is Not 25

What is the next number in the sequence?

1, 4, 9, 16, ...

Probably 99 out 100 students would give the expected answer of
25, without second thoughts to other possible valid answers. This is
simply because seldom do we point out to pattern-seekers that the
obvious answer is just one of many possible (infinite) answers, as
there are other rules (or expressions or functions) that can generate the
same sequence of numbers.

Other than the expected square number 25, another answer to the
above *sequence* (also called *series*) is the number 49 if we take the
rule to be

$$(n-1)(n-2)(n-3)(n-4) + n^2$$

instead of n^2.

Clearly, replacing $n = 1, 2, 3$, and 4 in both expressions yields the first
four terms (1, 4, 9, 16).

The General Term

In general, if the first few terms of a given sequence are a, b, c, how
do we then formulate a rule if we want the next number to be some
value x, where x is some real number?

First form the expression

$$(n-1)(n-2)(n-3)$$

which will yield the value 0 for $n = 1, 2$, and 3.

Then add to $(n-1)(n-2)(n-3)$ some expression in n that produces a, b, and c when 1, 2, and 3 are substituted for n.

Thus the 4th term produced by $(n-1)(n-2)(n-3)$ plus some expression is 6 more than that expression, since $(4-1)(4-2)(4-3) = 3 \times 2 \times 1 = 6$.

Now, if the expression in n that gives a, b, and c gives d as the 4th term, then we will have $d + 6$. This means that when $n = 4$,

$$\text{expression in } n = d$$

and

$$(n-1)(n-2)(n-3) + \text{expression in } n = 6 + d$$

However, $d + 6$ needn't necessarily be our desired number x as the next term.

So, to get x, we multiply the expression $(n-1)(n-2)(n-3)$ by some value p such that

$$p(n-1)(n-2)(n-3) + d = x$$

or

$$6p + d = x$$

On solving, $p = \frac{x-d}{6}$.

> The desired next term is x when the expected answer is d.

For example, if we want to generate the sequence

$$1, 2, 3, 8$$

then $x = 8$, $d = 4$; thus $p = \frac{x-d}{6} = \frac{8-4}{6} = \frac{2}{3}$.

Hence the sequence 1, 2, 3, 8 can be obtained by substituting 1, 2, 3, 4 in order in the formula or expression

$$\tfrac{2}{3}(n-1)(n-2)(n-3) + n$$

To impress our audience, we can multiply the above expression to yield

$$\tfrac{2}{3}n^3 - 4n^2 + \tfrac{25}{3}n - 4$$

which is the correct expression to give

$$1, 2, 3, 8$$

when 1, 2, 3, 4 is substituted for n in order.

When the Next Term Is π

Let's work out the rule or formula for the following sequence

$$1, 2, 3, \pi$$

Here, $x = \pi$, $d = 4$; so $\frac{\pi - 4}{6}$.

Our desired next term is π; our expected answer is 4.

Hence the formula to generate the above sequence with the number π as the 4th term is given by

$$\frac{\pi - 4}{6}(n - 1)(n - 2)(n - 3) + 4$$

Check: $n = 4$, $\frac{\pi - 4}{6}(4 - 1)(4 - 2)(4 - 3) + 4 = \frac{\pi - 4}{6} \times 3 \times 2 \times 1 + 4 = \pi$

In general, given a sequence of numbers, we can always find the rule which will yield a desired number as our next term. In other words, finding the next term of a given sequence has not just one, but an infinite number of equally valid answers, depending on the rule used to generate each sequence.

This is why questions asking the next term of a sequence or series seldom appears in examinations, unless they are set in a multiple-choice format. However, these number sequences do pop up frequently in IQ or aptitude tests to puzzle the problem solvers.

Practice

Find the general term (or rule) which will generate the following sequences.

(a) 1, 2, 3, **5**
(b) 1, 4, 9, 16, **40**
(c) 1, 2, 3, 5, **4**

Christmas Countdown

. . . .

Dec 27 Buy Christmas cards and ornaments and wrapping paper for next year at half price, during post-Xmas sales.

Feb 5 **Buy your first Christmas present while visiting Batam, Indonesia.**

May 14 Buy your second Christmas present from door-to-door ex-convict salesman you feel sorry for.

Nov 28 **Buy more cards and ornaments and wrapping because you can't remember where you put the ones you bought last year.**

Dec 7 Spend the day in bed browsing through all the catalogues, wishing you could afford the nicer things, or hoping that someone will splurge on you.

Dec 14 **Set to go on a diet right after Christmas (or maybe the New Year). Better, discipline yourself for a Joseph-like fast.**

Dec 15 Complain about the over-commercialization of Christmas, and the paganization of the birthday of Jesus Christ.

Dec 18 **Buy a Christmas tree that costs what your grandfather got for a weekly salary when he first started working.**

Dec 22 Send cards to the people you forgot about.

Dec 23 **Buy nothing, reluctantly supporting the "Buy Nothing Day" campaign.**

BUY
NOTHING
DAY

Dec 25 Smile bravely and try to say, "Thank you" as if you really meant it. Experience mixed feelings of not being able to buy more expensive gifts for colleagues and friends, and of being disappointed at receiving inexpensive (probably from last year's) Xmas gifts from clients and relatives.

Dec 26 **Oversleep and too tired to even unwrap the Christmas gifts, following the previous day's Xmas wild party.**

Dec 27 The cycle repeats itself *ad infinitum*.

Why do mathematicians often confuse Christmas and Halloween?

Because Oct. 31 = Dec. 25.

A Christmas Potpourri

1. A Joyful Ride

There are 25 people lining up for an amusement park ride. The ride can take up to 7 people at a time. How many times will the ride have to go for all 25 people to go on the ride once?

2. A Christmas Party

At a Christmas party, 25 guests shook hands with each other only once. How many handshakes were exchanged altogether?

3. Christmas Letters

The following letters have something in common, except two.

C H R I S T M A S

Which letters do not belong to the group? Why?

4. Time for Delivery Gifts

Santa and five volunteers can deliver 6 gifts in 6 minutes. How many more volunteers does he need to deliver 25 gifts in 25 minutes?

5. Rapid Computation

Without the use of a calculator, or of pencil and paper, how can you mentally calculate the sum of the numbers from 1 to 25 inclusive?

6. The Angle

What is the angle between the hands of a clock at 1:25?

7. Writing 25

What is another way of getting the number 25 using the numbers 2 and 5 only once, other than equating it to 5^2?

8. A Christmas Star

How many triangles and diamonds?

How many kites and pentagons?

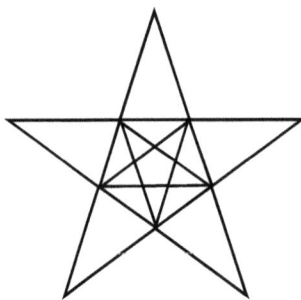

(a) How many triangles, diamonds, kites, and pentagons are there?
(b) What other shapes can you find?

9. Christmas Trees

You plant a small Christmas tree in your garden. In the first year it grows one stem. From then onwards it doubles the number of stems each year, and each stem grows two fir cones. How many stems and fir cones have grown after 1, 2, 3, ..., 25 years?

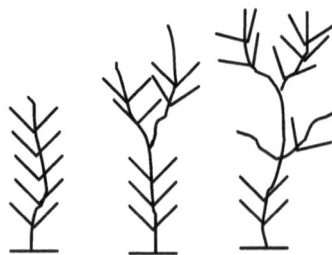

10. Time Needed

Determine how long it will take to return all the gifts mentioned in the song "The twelve days of Christmas" if the gifts are returned at the rate of one gift per day.

11. A Xmas Family

A mother who was born on December 25, married to a husband also born on December 25, gives birth to their first baby on December 25. What are the odds of the mother doing that?

12. Four Fours to Make 25

Using four fours, the four basic operations and, if necessary, and/or 4!, form the number 25. [Note: $4! = 1 \times 2 \times 3 \times 4$]

13. A Secret Message

What is the secret Christmas greetings encoded in the cryptic message?

$$\frac{C\ C\ C\ C\ C\ C}{\text{HELLO!\ \ HELLO!}}$$

14. The Unknown Digit

All the whole numbers beginning with 1 are written successively as

$$12345678910111213\ldots$$

Which digit occupies the 2525th position?

15. 25 days, 25 Hours, 25 Minutes, …

It is now 1:25 p.m. What time was it 25 days, 25 hours, and 25 minutes ago?

16. A Christmas Call

"Hi, is this 9001-3267?"
"Yes, who's on the line?"
"Don't you recognize my voice?"
My mother is your mother's mother-in-law!"

Who is on the phone?

17. Santa vs. Satan

Prove that Oct. 31 = Dec. 25.

18. A Recurring Decimal

Show that $0.039999\ldots$ is exactly equal to $\dfrac{1}{25}$.

19. A Guesstimation

Estimate the value of $(1.01)^{25}$, giving your answer to 10 significant figures.

20. Who Is Taller?

A general is choosing a guard from 625 soldiers. He orders them to form a 25×25 square. He orders the tallest man in each row to step aside and chooses the shortest of the twenty-five. Then he changes his mind and has them go back to their places. He orders the shortest man in each column to step aside and chooses the tallest of these twenty-five. The two guards chosen by the two methods are different. Which one is taller?

21. Distinct Neighbors

There are 25 guests sitting round a big circular table. How many of the 25! arrangements are distinct, that is, do not have the same neighboring relations?

$25! = 1 \times 2 \times 3 \times \cdots \times 24 \times 25$

22. A Millionairess on Retirement

Ruth earns an annual salary of $25,000 in the first year. She will receive an annual salary increase of 2.5% each year. Assuming she works for the company until she retires, calculate
(a) the amount she will earn in her 25th year,
(b) the total amount she will earn over a 25-year period,
(c) the least number of years she has to work before her earnings exceeds one million dollars.

23. Four Nines

Using only four nines, any of the four basic operations $(+, -, \times, \div)$, parentheses and, if necessary, $\sqrt{9} = 3$ and $.\overline{9} = 1$ form the number 25.

24. Twenty-five Images of Zero

Using only four nines, the four basic operations and, if necessary, $\sqrt{9}$ and/or $.\overline{9} = 1$, write 25 representations for the number zero.

25. A Christmas Party

At a Christmas party, each child brought a present. Presents were put in a large basket. All presents were different but identically wrapped. Going home, each child randomly selected a present from the basket. What is the expected number of children who carry home their own presents?

Dear Santa

I wish I was Mary Christmas!

A Question of Inequality
The 25 cents meal
and the $50 buffet

Political
Correctness

The next Christmas Day solar eclipse won't happen until 2307.

The Red Cross

↓

The Green Cross

Be Santa's second-in-command
Taking the pictures with Santa,
please line up...

CHRISTMAS
Alphametics

. . . .

Cryptarithms are puzzles in which letters or symbols are replaced by the digits in an arithmetical calculation. If a cryptarithm uses letters in place of the digits, and these letters form sensible words or phrases, the puzzle is termed an *alphametic*, a term coined by J. A. H. Hunter in 1955.

Let's enjoy cracking the codes of some alphametics with a Christmas flavor. In each case, replace each letter with a digit, different letters being different digits.

1. Christmas Greetings

(a)
$$A \overline{)\ \frac{XMAS}{MERRY}}$$

(b)
$$A \overline{)\ \frac{XMAS}{HAPPY}}$$

(a) has four solutions.
(b) has two solutions.

2. More Xmas Greetings

(a)
```
        A
    MERRY
+   XMAS
  TURKEY
```

(b)
```
  SANTA
− CLAUS
  XMAS
```

(a) has one solution.
(b) has two solutions.

3. Maximum Number of Days in a Year

DAY + DAY +⋯+ DAY = YEAR, where the number of addends is a maximum. (Brian Barwell)

4. Quadrennially by Los Acertijeros

366 + DAYS = LEAP + YEAR, where YEAR is a maximum and 3 and 6 cannot be reused.

5. Old Testament by Randall J. Covill

ADAM + EVE + CAIN + ABEL = BIBLE, where BIBLE is a maximum.

6. Adam & Eve

$$
\begin{array}{r}
ADAM \\
AND \\
EVE \\
ON \\
+\quad A \\
\hline
RAFT
\end{array}
$$

7. Joyeux Noel

$$
\begin{array}{r}
NOEL \\
+\ NOEL \\
\hline
BELLS
\end{array}
$$

8. Talkative Eve

$\dfrac{EVE}{DID}$ =.TALKTALKTALK… represents a common fraction expressed as an unending decimal. What digits do the letters stand for?

Hint: The solution is unique.

> The fraction EVE/DID is in its lowest terms.

9. Yuletide Sentiment by Harry L. Nelson

Solve the alphametic

(MERRY) (XMAS) = CHRISTMAS, where CHRISTMAS is maximal in base 13.

10. Fewer Toys, More Cash

$$
\begin{array}{r}
\text{SAVE} \\
+ \text{MORE} \\
\hline
\text{MONEY}
\end{array}
$$

Hint: Four solutions are possible.

11. More Happy Days to Come

$$
\begin{array}{r}
\text{HAPPY} \\
\text{HAPPY} \\
\text{HAPPY} \\
+ \quad \text{DAYS} \\
\hline
\text{AHEAD}
\end{array}
$$

12. Xmas Mail

$$
\begin{array}{r}
\text{XMAS} \\
\text{MAIL} \\
+ \text{EARLY} \\
\hline
\text{PLEASE}
\end{array}
$$

13. 2597 Garden Party
by A. G. Bradbury

ADAM + & + EVE + HAD + TEA + AT = EIGHT, where TEA is prime.

14. A Love Equation
by Alan Wayne

$$\sqrt{(PASSION)} = KISS$$

An ill-prepared college student taking a math exam just before Christmas vacation wrote on his paper, "Only God knows the answers to these questions. Merry Christmas!"
The professor graded the papers and wrote this note: "God gets 100, you get 0. Happy New Year!"

(Eves, 1988, pp. 149-150)

Mathematical Graphiti

Joseph was Jesus' father.
Jesus' father was God.
So Joseph was God!

The fool is one who keeps GOD out of the eternal equation.

Gift giving would have destroyed at least one-third of the value of the items transferred as gifts.

If Santa were a woman...

Adopt a turkey this Xmas!
– SPCA

The Buy Nothing @ Xmas Campaign

GIFTS

XMA$ $AVING$
• Buy Christmas candy on December 26
• Use recycled calendars
• Plan your Xmas Bahamas holidays when the tourists are leaving.

Act: For terrorizing innocent children
Reward: 25 Santas (*alias* Satan) are waiting for you in Hades!

25 Not-to-do Lists

Heavenly Communiqué
Jesus would like to distance Himself from Santa.

❈ For *Christmas* church goers Reward: Temperature in hell will cool down by 25.25°F every quarter century ❈

TGIC – Thank God It's Christmas

WWJS

Satan ~~Santa~~'s Gift for Dad on Xmas: cigarettes & liquor

Celebrate
Father Christmas Week

To make your Christmas memorable, many creative ways exist to celebrate the most popular holiday in the entire calendar, other than sending Christmas cards or giving presents to family members, friends, and colleagues.

Here are ten easy and meaningful ways you could help to play your part during the festive week.

1. **Unsolicited Greetings**
   From a phonebook, randomly pick up some names and call these persons to wish them a blessed Christmas and prosperous New Year.

2. **Unexpected Xmas Gifts**
   Surprise some of your neighbors who would least expect any gifts from you. You could have these presents or gift vouchers delivered to them anonymously.

3. **Personalized Xmas Cards**
   Design a personal Christmas card and pose a dozen CHRISTmaths crackers for your recipients to solve.

> 12. *What is the sum of the interior angles of the Christmas tree?*
>
> 11. *What day of the week is Christmas Day in 2025? ...*
> $\vdots$
> 1. *How many terminal zeros are there in* $1 \times 2 \times 3 \times \cdots \times 24 \times 25?$

Remember to include the solutions; at the least, give idiot-proof hints.

4. **A Father Santa's Check**
   Mail a surprise check to some neighbors or friends-in-need across the street or block, without identifying yourself. You can say, "It's a gift from your Father in Heaven, and you happen to be a God-sent postman!".

5. **A Christmas Grace**
   Offer to bail a stranger out of jail. Besides settling their fines, bless them with some pocket money, too! Make it their most memorable Christmas ever!

6. **CA$H on a Christmas Tree**
   Hide some $100 notes among some trees in your neighborhood. Give dummy-proof instructions for your recipients to locate and collect them.

7. **Christmas on Christmas Island**
   Donate two free tickets to a couple about to celebrate their silver jubilee for a weeklong holiday trip to Christmas Island, with hotel accommodation expenses fully paid.

8. **Shopping with Santa**
   Enter a shop and pay for the items picked up by a shopper before he or she makes his or her way to the cashier.

9. **Adopt a Turkey for Life.**
   Get a certificate that you are the sole owner of the lucky bird. If it is a "she," you may donate her eggs for research purposes.

10. **Facebookers' Party**
    Invite 25 new friends to join you at a Xmas party, all expenses paid, to hear the true meaning of the Christmas story.

25 Illegal Things You May Want to Do on Xmas (at your own risk)

. . . .

25. **Carve "Merry CHRISTmaths" on a tree.**

24. Sell Möbius decorations for a Christmas tree without a licence.

23. **Visit a cemetery and place flowers on an old unattended gravestone—for someone who died on December 25.**

22. Send roses to your fiends or foes on Xmas Day (business foes, ex-employees you had fired, ex-partners you had sued, …) with best wishes from you.

21. **Mail a copy of *CHRISTmaths* to ex-classmates or ex-colleagues who take pride in their mathematical ineptitude; tell them you are praying for their innumeracy to evaporate away.**

20. If a suitor calls you, do not answer the phone before the 25th ring. Make him think you are busy even if you are doing nothing.

19. **Put Xmas graffiti on a cow; however, avoid cow-worshipping places at all costs—you have been duly warned about "Stoning by Death" or "Death by Hostile Mobs."**

18. If you are a woman, dress up as Santa Claus on city streets.

17. **Mow a crop circle of Santa, or some depictions associated with Christmas.**

16. Stick a Christmas tree on some graves and light them with candles to "invite the corpses in the celebrations."

15. **Conduct a random phone survey of residents in your area, and report the average holiday spending for 25 household incomes. Then depict the common shopping items purchased on a pie chart.**

14. Pick up the phone book and wish "Merry Xmas" to 25 people with the first or last name, "Christmas."

13. **Erect a Santa-on-the-cross in your garden or yard to protest the commercialization during the run-up to Christmas.**

12. Avoid stores that prefer the politically correct "Happy Holidays" instead of "Merry Christmas" in their advertising and store decorations.

11. **Set up a Website of stores that use the phrase "Happy Holidays" along with a poll that asks, "Will you shop at stores that do not say 'Merry Christmas'?"**

10. Share the Good News to open-minded atheists and agnostics, who are receptive to hearing the gospel of Jesus Christ.

9. **Interview someone born on Christmas Day; better still, whose name is Christmas, and send your story to a local newspaper to be considered for publication.**

8. Persuade your neighbors to support a "Buy Nothing at Christmas" campaign; instead, channel the money or bonus to some developing countries to help the poor and underprivileged.

7. **Provide Christmas gifts to children with a non-Christian parent in prison. Adopt a boy or girl, learn his or her age and interests, then go Christmas shopping for that child.**

6. Send beautifully wrapped boxes to strangers with nothing inside—just the pure spirit of giving "zero gifts."

5. Put an ad in your local paper which informs readers that toys are made affordable for millions of parents, thanks to the appalling working conditions endured by under-aged children in Santa's sweatshops, in China.

4. Work out and promote the plan of a "25 Hour Protection Against Innumeracy" to insurance companies, and earn a 2.5 percent commission.

3. Have beauty parlors offer a "How to Look 25 Years Younger" package to their customers, by letting the Photoshop lie for them.

2. Collect and sell retails- and factory-reject boxes, to be used as Christmas decorations.

1. Select the best attractive offer from the first twenty-five buyers keen on buying your soul online.

A Green Christmas

The biggest victims during the festive season are not lonely singles or poor and hungry people but trees—Christmas trees. Unless we declare war on trees, we should not expect less carbon dioxide in the atmosphere.

How much deforestation would result each year if people chopped down their trees from a forest rather than getting an artificial tree or getting one from a tree farm?

Is it less costly to buy an artificial Xmas tree than one cut down from the forest or tree farm in the long run?

If 10 percent of the 3.0×10^8 people in the United States were to buy a Christmas tree every year, this would translate into 30 million trees.

Since these trees come in all sizes and shapes, let's assume that in a forest, on average, they are three meters apart, so each tree would occupy an area of $3 \times 3 = 9$ square meters, or 10 square meters to round off to the nearest power of ten.

The area of deforestation may be calculated as follows:

$$\text{Deforested area} = \text{number of trees} \times \text{area per tree}$$
$$= (3.0 \times 10^7 \text{ trees}) \times (10 \text{ m}^2/\text{tree})$$
$$= 3.0 \times 10^8 \text{ m}^2$$
$$= 3.0 \times 10^2 \text{ km}^2$$

So every Christmas, about 300 km² of deforestation take place in the U.S.

Run a Christmas tree farm, today!

To profit from a Christmas tree farm, start now —you'll reap the rewards in around four to fifteen years.

Green Business

1. How much carbon dioxide would be sulked out of the atmosphere every year if all nations agreed to sign a global treaty to use artificial instead of forest or farm trees for Christmas?

2. **If you decide to invest in a tree farm how much could you expect to earn in a quarter century?**

3. From an economic and ecological standpoint, would it better to use a farm or forest tree rather than an artificial tree? How many pounds of carbon dioxide could be saved in the long run?

4. **If you are a *Christrepreneur* committed to playing your part in reducing global warming, what innovative products could you come up with in making the festive season more eco-friendly?**

5. Suggest 25 cool ways how math educators could go green, without compromising on educational and ethical standards.

The Twelve Days of Christmas

The popular Christmas song "The Twelve Days of Christmas" serves as a very good enrichment mathematics topic to promote critical thinking among students for the holidays. Problems on sequences, series, and patterns would help motivate students to come up with clever ways to solve them, while enhancing their higher-order thinking skills.

The first day of Christmas
My true love sent to me
A partridge in a pear tree.

The second day of Christmas
My true love sent to me
Two turtle doves, and
a partridge in a pear tree.

The third day of Christmas
My true love sent to me
Three French hens, two
turtledoves and a partridge
in a pear tree.

The fourth day of Christmas
My true love sent to me
Four calling birds, three French
hens, two turtledoves and a
partridge in a pear tree.

The fifth day of Christmas
My true love sent to me
Five gold rings, four calling birds,
three French hens, two turtledoves
and a partridge in a pear tree.

The sixth day of Christmas
My true love sent to me
Six geese a-laying, five …

The seventh day of Christmas
My true love sent to me
Seven swans a-swimming, six ...

The eighth day of Christmas
My true love sent to me
Eight maids a-milking, seven ...

The ninth day of Christmas
My true love sent to me
Nine drummers drumming, eight ...

The tenth day of Christmas
My true love sent to me
Ten pipers piping, nine ...

The eleventh day of Christmas
My true love sent to me
Eleven ladies dancing, ten ...

The twelfth day of Christmas
My true love sent to me
Twelve lords a-leaping, eleven ...

(a) How many gifts did my true love give me on the first day? On the third day? On the seventh day? On the nth day?

(b) What is the total number of gifts did my true love give to me on the first day? On the third day? On the twelfth day?

(c) What is the total number of items did my true love give me on n days?

Solution

(a) On the first day, only one gift, the partridge in a pear tree, was given.

On the third day, six gifts were given: three French hens, two turtledoves, and a partridge in a pear tree.

On the seventh day, 28 gifts were given:

$$7 + 6 + 5 + 4 + 3 + 2 + 1 = 28$$

On the nth day, $1 + 2 + 3 + \cdots + n = \dfrac{n(n+1)}{2}$ gifts were given.

The number of gifts given on the nth day is equal to the nth triangular number.

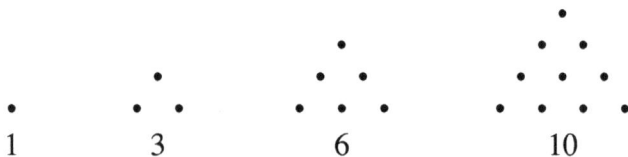

$$\quad 1 \qquad\qquad 3 \qquad\qquad 6 \qquad\qquad 10$$

(b)

Day	Number of items given on that day	Number given on all days combined
1	1	1
2	3	4
3	6	10
4	10	20
5	15	35
6	21	56
7	28	84
8	36	120
9	45	165
10	55	220
11	66	286
12	78	364

On the third day, a total of $1 + 3 + 6 = 10$ gifts were given.

The table shows that 364 items were given on the twelve days.

The total is the sum of the first twelve triangular numbers:

$$1 + 3 + 6 + 10 + 15 + 21 + 28 + 36 + 45 + 55 + 66 + 78 = 364$$

In combinatorial terms, it can be shown that the number of the first n triangular numbers is given by

$$\binom{n+2}{3} = \frac{1}{6}(n^3 + 3n^2 + 2n)$$

The Twelve Days of Christmas and Pascal's Triangle

It is not difficult to see that the daily collections of gifts are 1, 4, 10, 20, ..., which are the pyramid numbers found in Pascal's triangle.

Pyramid numbers are a three-dimensional equivalent of the triangular numbers. In other words, a stack of triangular numbers yield the pyramid numbers, as depicted below.

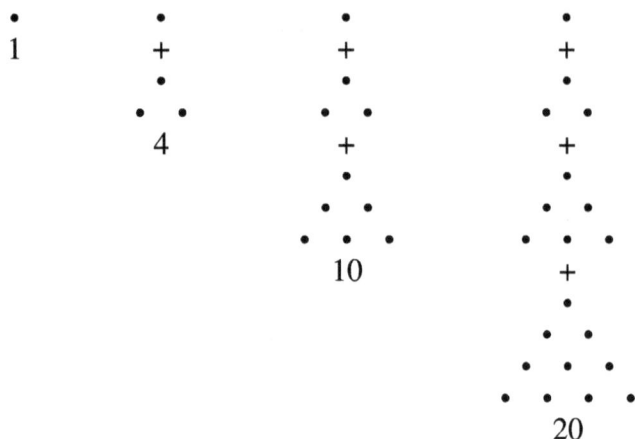

Can you find the triangular and pyramid numbers in the Pascal's triangle? Let's consider the first five rows of the Pascal's triangle.

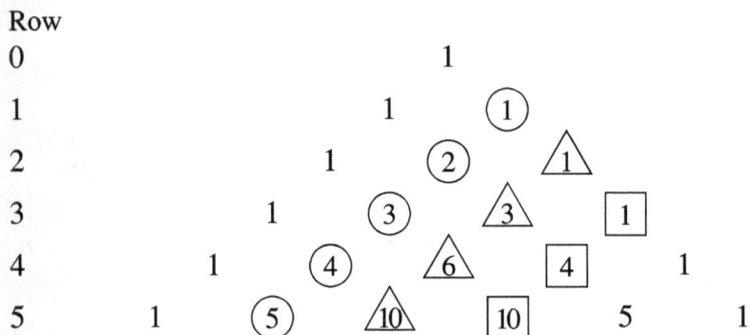

To find the number of gifts received on the 12th day, extend the triangle to the 13th row, go over three places, and read 78. Verify that this is so. How would you use the Pascal's triangle to find the total number of gifts collected by the end of the 12th day?

The Christmas Cubic

Earlier on, we saw that the number of gifts, g, received on the nth day is given by the formula $g = \binom{n+2}{3} = \frac{n(n+1)}{2}$. Let's use the method of finite differences to find a formula for the total number of gifts collected on the nth day.

The total numbers of gifts received in the first few days are shown in the first row below. After three subtractions, the difference between the values is constant, which implies that the sequence of the numbers in the first row is generated by a cubic polynomial.

$$
\begin{array}{ccccccccc}
1 & & 4 & & 10 & & 20 & & 35 \\
& 3 & & 6 & & 10 & & 15 & \\
& & 3 & & 4 & & 5 & & \\
& & & 1 & & 1 & & &
\end{array}
$$

The general cubic polynomial can be written as $ax^3 + bx^2 + cx + d$, where x denotes the number of days.

If $x = 1$, the number of gifts is $a(1)^3 + b(1)^2 + c(1) + d$, i.e., $a + b + c + d = 1$
If $x = 2$, the number of gifts is $a(2)^3 + b(2)^2 + c(2) + d$, i.e., $8a + 4b + 2c + d = 4$
If $x = 3$, the number of gifts is $a(3)^3 + b(3)^2 + c(3) + d$, i.e., $27a + 9b + 3c + d = 10$
If $x = 4$, the number of gifts is $a(4)^3 + b(4)^2 + c(4) + d$, i.e., $64a + 16b + 4c + d = 20$

Solving the four equations simultaneously, we have

$$a = \frac{1}{6}, b = \frac{1}{2}, c = \frac{1}{3} \text{ and } d = 0.$$

So the cubic expression is $\dfrac{x^3}{6} + \dfrac{x^2}{6} + \dfrac{x}{3}$.

Use as many strategies as possible to solve the following nonroutine questions, which are all based on "The Twelve Days of Christmas."

1. Show that the number of gold rings was the same as the number of maids a-milking.

2. After 12 days of Christmas, which species of bird has the largest number?

3. How many birds were received in the traditional Christmas carol "The Twelve Days of Christmas"?

4. Five gold rings are given for the first time on the fifth day.
   (a) Show that five gold rings were given on eight days.
   (b) Formulate an expression that yields the total number of items brought throughout the song if n items were given for the first time on the nth day.

5. Use the Pascal's triangle to find the number of gifts received on the 12th day of Christmas, and the total number of gifts collected by the end of the 12th day.

6. Imagine your beloved has deep pockets and he or she could buy you all the items listed in the popular Christmas jingle "The Twelve Days of Christmas."
   (a) How much would the collection of items cost? How much cheaper would it be if you decided to purchase them over the Internet?
   (b) Would $100,000 be sufficient to pay for all the gifts? What about $500,000?

> *A partridge in a pear tree: $210.00*
> *Two turtle doves: $95.50*
> $\vdots$
> *11 pipers piping: $2.796.75*
> *12 drummers drumming: $3,715.00*
> $\vdots$

7. How many diagonals are there in a dodecagon, a polygon with twelve sides?

8. Given that the number of diagonals in a polygon is generated by a quadratic polynomial, $ax^2 + bx + c$, where x is the number of sides of the polygon, use the method of finite differences to show that the number of diagonals in a polygon is $\frac{x(x-3)}{2}$, where x is the number of sides.

9. Show that the total number of gifts, t, received on the nth day is given by
$$t = \binom{n+2}{3}$$

10. Generalize "The Twelve Days of Christmas" to "The One Hundred Days of Christmas."

CHRISTmaths
ONLINE

For more *quickies*, *trickies*, and *toughies*,
visit the facebook page: **fb.com/christmaths**

Answers/Hints/Solutions

B.C. and A.D (p. 8)

0. How could the Romans know, when they supposedly coined that piece of money in "53 B.C.," that Christ would be born 53 years later? (*Quantum* brainteaser, Mar/Apr 1993, #3)

1. 1174 years. No year is numbered zero.

2. A birthday is the day one was born. One might have celebrated it ten times over ten years, but one still had only one birthday, namely the day one was actually born.

3. 99. The year 1 B.C. was the year before the birth of Christ. There was no year 0. The year 1 A.D. began with the birth of Christ.

4. Zero dollars. The coin was counterfeit because the term "B.C." could not have been used then. "B.C." means "before Christ," and the minter could not know Christ was going to be born 156 years after he minted the coin.

5. Of all the single-digit numbers, eight requires the most letters: VIII. We should use as many 8's as possible. The number is 1888, written as "MDCCCLXXXVIII."

6. 93.

7. 1 B.C.

There were 29 (or 129 – 100) years between Cleopatra's death and Boadicea's birth, so the answer is 1 B.C., 29 years after 30 B.C. Draw a number line to represent the lifespans.

8. The man was born in 1180 BC and died in 1163 B.C. (17 years later.)

9. 554 B.C.

10. 427 B.C.

11. 75 years.

12. A.D. 1642.

13. 1.

14. B. [5300 – 2000 = 3300]

15. Answers vary.

16. (a) Year –20.
(1965 – 1945)

17. January 1, 1901.
The first century A.D. began on January 1, 1.
On December 31, 99, ninety-nine years of the first century had elapsed. To complete the first century, we must add the entire year 100, ending on December 31, 100. Likewise, the 19th century ended on December, 1990.

18. The third millennium starts on 1 January 2001, not 1 January 2000.

The 12 Puzzles of Christmas (p. 15)

1.

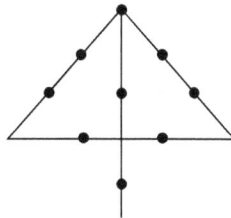

2. 0 angels.

5 persons can deliver 5 toys in 5 minutes.

5 persons can deliver 100 toys in 100 minutes.

So Santa and his four helpers can do the job.

Hence, he does not need any more angels.

3. Monday—calendars may be reused every 28 years.

4. The letter "l" is missing. The word is NOEL (No l).

5. 35 triangles.

6. Using the code A = 1, B = 2, and so on, "Merry Christmas" is 189.
A "Happy New Year" is 158.

7. There are only four spelling mistakes. But there is a fifth mistake: the claim that there are eight letters in the word "Christmas"!

8.

9. Reducing the common letters yields CHRIST = LOVE.

10.

11. 11 squares.

5 small   4 medium   2 large

12. 364 gifts.

The total number of gifts is $1 + (1 + 2) + (1 + 2 + 3) + \ldots + (1 + 2 + 3 + \ldots + 12) = 364$.

Item	No of items given	No of days given	Total no of each item
Partridge in a pear tree	1	12	12
Turtledoves	2	11	22
French Hens	3	10	30
Calling birds	4	9	36
Gold rings	5	8	40
Geese a-laying	6	7	42
Swans a-swimming	7	6	42
Maids a-milking	8	5	40
Ladies dancing	9	4	36
Lords a-leaping	10	3	30
Pipers piping	11	2	22
Drummers drumming	12	1	12

Santa's Itinerary *(p. 20)*

1. 6 possible tracks.

In combinatorial terms, the answer is given by $^4C_2 = \dfrac{4 \times 3}{1 \times 2} = 6$.

2. 13 ways.

3. 70 routes.

```
1   5   15   35   70
1   4   10   20   35
1   3    6   10   15
1   2    3    4    5
1   1    1    1    1
```

Alternatively,
Look for a possible pattern:
    2×2 yields 6 routes.
    3×3 yields 20 routes.
    4×4 yields 70 routes.

4. 252 routes.

Every number in the following array represents the number of shortest routes from point X to the corresponding corner point in the grid.

```
X
  1   1   1   1    1     1
  1   2   3   4    5     6
  1   3   6  10   15    21
  1   4  10  20   35    56
  1   5  15  35   70   126
  1   6  21  56  126   252
                            Y
```

The total number of shortest paths from X to every other corner point on a 5×5 grid map (determined by counting eastern moves) is given here in terms of the numbers $\binom{n}{r}$.

In combinatorial terms, the answer is given by $^{10}C_5 = \dfrac{10 \times 9 \times 8 \times 7 \times 6}{1 \times 2 \times 3 \times 4 \times 5}$

$= 252.$

5. 12 different paths.

6. 48 ways.

7. 80 ways.

8. 35 policemen.

9. 60 routes.

Alternatively,

There are 10 routes from A to X.
There are 6 routes from Y to B.
Hence there are $10 \times 6 = 60$ routes from A to B.

10. 784 routes.

11. 210 routes.

There are more than one shortest path between two points. For example, to go to a point halfway around a block, one can move clockwise or counterclockwise—both ways are equally short.

Observe that if the plan of the blocks were superimposed on Pascal's triangle, point X is located at the point marked 210. Thus, there are 210 equally short paths between point X and point Y.

12. 52 ways.

Total number of ways from P to Q = 26 + 26 = 52

A Christmas Spell *(p. 28)*

1. (a) $1 \times 2 \times 3 \times 4 \times 5 \times 4 \times 3 \times 2 \times 1$
    $= 2,880$
   (b) $1 \times 3 \times 5 \times 7 \times 9 \times 7 \times 5 \times 3 \times 1$
    $= 99,225$

2. 20 ways.

3. 144 ways (= $1 \times 2 \times 3 \times 4 \times 3 \times 2 \times 1$).

4.

3-step	4-step	5-step
CHIT	CHRIT	CHRIST
CRST	CHRST	CHRSIT
CRIT	CHIST	CHRSIT
	CRIST	CHIRST
	CRSIT	CRHIST
	CRHIT	

5. 6 ways.

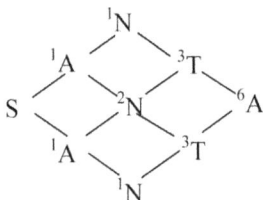

6. 20 ways.

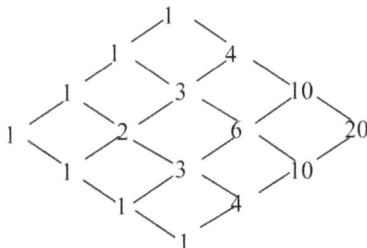

12 Challenges @ Christmastime (p. 38)

1. 5.

2. 180°.

Interior angle of a regular pentagon

$$= 180° - \frac{360°}{5}$$
$$= 108°$$

In isosceles $\triangle AEB$, the base angles

are $\dfrac{180° - 108°}{2} = \dfrac{72°}{2} = 36°$

So $\angle EAD = \angle BAC$
$\qquad = \angle x$
$\angle DAC = 108° - 2\angle x$
$\qquad = 108° - 72°$
$\qquad = 36°$

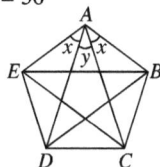

3. 1060 pages.

Pages	Number of digits
1–9	9
10–99	$90 \times 2 = 180$
100–999	$900\ 3 = 2700$
1000–?	

Number of digits used for 4-digit
pages $= 3133 - (9 + 180 + 2700)$
$\qquad\qquad = 244$
Number of 4-digit pages $= 244 \div 4$
$\qquad\qquad\qquad = 61$
So the number of pages the almanac
has $= 1000 + 61 - 1 = 1060$

4. Yes.

5. arc, sum, one, two, ray, pair, day, line, sphere.

6. About 5.52.

Let $x = \sqrt{25 + \sqrt{25 + \sqrt{25 + \cdots}}}$

Then $\qquad x = \sqrt{25 + x}$
$\qquad\qquad x^2 = 25 + x$
$\qquad x^2 - x - 25 = 0$
$$x = \frac{1 \pm \sqrt{101}}{2}$$

Since $x > 0$, $x = \dfrac{1 \pm \sqrt{101}}{2} = 5.52$ (to 2 decimal places)

7. 40 local cards, 17 letters, 13 overseas cards, or 10 local cards, 28 letters, 12 overseas cards. If we accept no mail of a particular type, there is an extra answer of 0 local cards, 39 letters, and 11 overseas cards.

Let the number of local cards, letters and overseas cards be a, b, and c respectively.

Then $\qquad a + b + c = 50 \qquad\qquad (1)$
and
$\qquad 40a + 45b + 95c = 2800 \quad (2)$

Dividing equation (2) by 5:
$\qquad 8a + 9b + 19c = 560$
Multiply equation (1) by 8:
$\qquad 8a + 8b + 8c = 400$
Subtracting: $b + 11c = 160$
$\qquad\qquad\qquad c = (160 - b)/11$

Also, $c < b$

b	6	17	28	39	50	...
c	14	13	12	11	10	...

If $b = 17$, $c = 13$, $a = 20$ [17 > 13]
Check: $40 \times 20 + (45 \times 17) + 95 \times 13)$
$= 2800$

If $b = 28$, $c = 12$, $a = 10$
Check: $40 \times 10 + (45 \times 28) + (95 \times 12) = 2800$

$b = 39$, $c = 11$, $a = 0$
Check: $39 \times 45 + (11 \times 95) = 2800$

8. 38 triangles (24 with 1-unit sides, 12 with 2-unit sides, 2 with 3-unit sides)
$24 = 5 + 7 + 7 + 5$; $12 = 6 + 6$

9. From the time that both were the same length, it took 2 hours for the longer to burn out, $1\frac{1}{2}$ hours for the shorter.

Thus the rate of burning of the longer is $\dfrac{1\frac{1}{2}}{2} = \dfrac{3}{2} \times \dfrac{1}{2} = \dfrac{3}{4}$ that of the shorter.

Let r be the rate of burning of the shorter candle; then $6 \times (3r/4) = 4r + 1$. Thus the rate of the shorter is 2 cm per hour, and the rate of the longer is 1 cm per hour.
So the longer was 9 cm and the shorter was 8 cm.

10. $\sqrt{x + \sqrt{x + \sqrt{x \cdots}}} = 25$
Square each side:
$x + \sqrt{x + \sqrt{x + \sqrt{x \cdots}}} = 625$
$x + 25 = 625$
$x = 600$

11. At midnight when the hands of the clock form a Christmas Tree.

12. The sequence repeats itself after every 9 letters.

$\underbrace{\text{MERRYXMAS}}_{\text{9 letters}} \underbrace{\text{MERRYXMAS}}_{\text{9 letters}} \ldots$

$2525 = 280 \times 9 + 5$

There are 280 groups of 9 letters followed by the first five letters of MERRYXMAS.

So the 2525th letter is Y.

Christmas Philamath (p. 45)

1. 4 tears.

2. 5 tears.

3. 6 tears.

4. 48 tears.

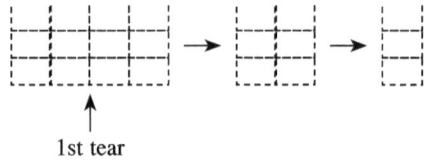

1st tear

6. 24.
Observe that each time we tear a piece of the sheet the total number of pieces increases by one. Since we need 25 pieces, we will have to tear the sheet 24 times.

7. There are $4 \times 3 \times 2 \times 1 = 24$ possible permutations, but only 16 possible folds.

8. Only 1¢ and 3¢ are the only amounts which cannot be made.

9. There are 24 permutations but only 8 possible folds:

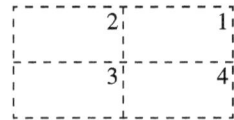

1234	2143	3214	4123
1432	2341	3412	4321

25 No-Frills Christmas Crackers (p. 54)

1. M, E, H.

2. True.

3. 300 handshakes.

$$(= 24 + 23 + \cdots + 2 + 1 = \frac{24 \times 25}{2}$$
$$= 12 \times 25 = 3 \times 4 \times 25)$$

4. $\underbrace{999\ldots99990.}_{24 \text{ 9s}}$

5. 100 cm — the perimeter is unaffected.

6. Friday.
   If tomorrow is Sunday, then today is Saturday, and yesterday was Friday.
   Now, $25 = 7 \times 3 + 4$
   25 days before yesterday
   = 4 days before yesterday
   = 4 days before Friday
   = Monday
   25 days after 25 days before yesterday
   = 25 days after Monday
   = 4 days after Monday
   = Friday

7. 5 cuts → 60 mins
   24 cuts → $\frac{60}{5} \times 24 = 288$ mins
   $\qquad\qquad = 2$ hrs 48 mins

8. Strictly speaking, the mathematicians are correct, as they want to differentiate between "minus 25" and "negative 25." However, in everyday language, "minus twenty-five" is commonly used among the public.

9. 252 routes.
   Every number in the following array represents the number of shortest routes from point A to the corresponding corner point in the grid.

1	1	1	1	1	1
1	2	3	4	5	6
1	3	6	10	15	21
1	4	10	20	35	56
1	5	15	35	70	126
1	6	21	56	126	252

In combinatorics, the total number of shortest paths from A to every other corner point on a 5×5 grid map (determined by counting eastern moves) is given here in terms of the numbers $\binom{n}{r}$.

In our case, the number of shortest paths is $\binom{10}{5} = \frac{10 \times 9 \times 8 \times 7 \times 6}{1 \times 2 \times 3 \times 4 \times 5} = 252$

10. The figure has 14 sides of 5 cm each

Perimeter of figure = 14×5 cm
$\qquad\qquad\qquad = 70$ cm
Area of one square = 5 cm × 5 cm
$\qquad\qquad\qquad = 25$ cm^2
Area of figure = 6×25 cm^2
$\qquad\qquad = 150$ cm^2

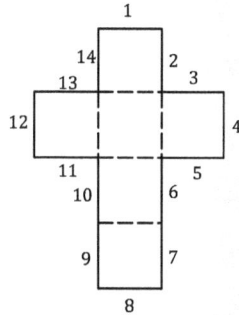

11. $\dfrac{1}{1 \times 2} = \dfrac{1}{2} = 1 - \dfrac{1}{2}$

$\dfrac{1}{2 \times 3} = \dfrac{1}{6} = \dfrac{1}{2} - \dfrac{1}{3}$

$\dfrac{1}{3 \times 4} = \dfrac{1}{12} = \dfrac{1}{3} - \dfrac{1}{4}$

$\dfrac{1}{24 \times 25} = \dfrac{1}{24} - \dfrac{1}{25}$

$\dfrac{1}{1 \times 2} + \dfrac{1}{2 \times 3} + \dfrac{1}{3 \times 4} + \cdots + \dfrac{1}{24 \times 25}$

$= 1 - \dfrac{1}{2} + \dfrac{1}{2} - \dfrac{1}{3} + \dfrac{1}{3} \cdots + \dfrac{1}{24} - \dfrac{1}{25}$

$= 1 - \dfrac{1}{25}$

$= \dfrac{24}{25}$

12. 24 snaps.

Observe that each time we snap a piece of the chocolate the total number of pieces increases by one.

Since we need 25 pieces, we will have to snap the bar 24 times.

13. Note that $\dfrac{1}{\sqrt{n}+\sqrt{n+1}} = \sqrt{n+1} - \sqrt{n}$

$\dfrac{1}{\sqrt{1}+\sqrt{2}} + \dfrac{1}{\sqrt{2}+\sqrt{3}} + \cdots + \dfrac{1}{\sqrt{24}+\sqrt{25}}$

$= \sqrt{2} - \sqrt{1} + \sqrt{3} - \sqrt{2} - \sqrt{4} - \sqrt{3}$

$+ \cdots + \sqrt{25} - \sqrt{24}$

$= \sqrt{25} - \sqrt{1}$

$= 5 - 1$

$= 4$

14. Friday.

15.

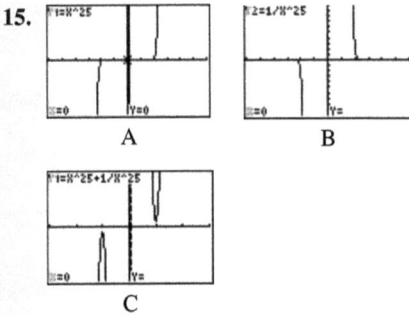

16. 624 snaps ($= 25^2 - 1$).

17. 3 ways.

$5	$2
5	0
3	5
1	10

18. HO HO HO ($3 \times H_2O - 2$).

19.

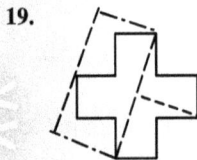

20. In 2.5 days Joseph makes a cross.

In 1 day Joseph makes $\dfrac{1}{2.5} = \dfrac{10}{25} = \dfrac{2}{5}$ of a cross.

In 4 days Mary makes a cross.

In 1 day Mary makes $\dfrac{1}{4}$ of a cross.

In 1 day Joseph and Mary make

$\dfrac{2}{5} + \dfrac{1}{4} = \dfrac{13}{20}$ of a cross.

So they will take $\dfrac{20}{13}$ days to make a cross.

Hence they will take $\dfrac{20}{13} \times 25 = \dfrac{500}{13}$

$= 38\dfrac{6}{13}$ days to make 25 crosses.

21.

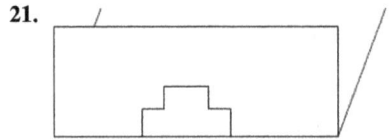

22. 100 cm².

If the side of each small square is x, then

$x^2 + (2x)^2 = 10^2$

$5x^2 = 10^2$

$x = 2\sqrt{5}$

Area of the cross $= 5 \times (2\sqrt{5})^2$

$= 100$ cm²

23. 49 contestants ($= 24 + 1 + 24$).

24. 21 squares.

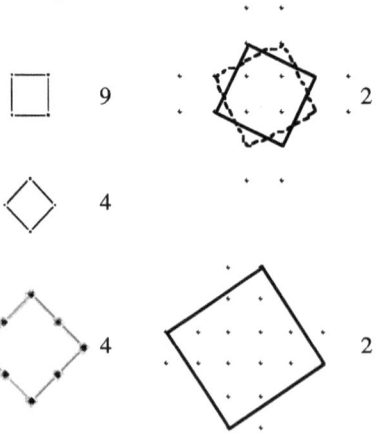

25. 75.

The first three values in the sum are 1, the next five are 2, the next seven are 3, the next nine are 4, and the final one is 5 for a total of

$3 \times 1 + 5 \times 2 + 7 \times 3 + 9 \times 4 + 5$
$= 75$.

The Mathematics of Christmas (p. 62)

0. $A = 4\pi r^2 = 4 \times \pi \times 6,400^2 \text{ km}^2 =$
$514,718,540.4 \text{ km}^2 \approx 510,000,000 \text{ km}^2$
Land surface area = 29% of
$510,000,000 \text{ km}^2 \approx 150,000,000 \text{ km}^2$

1. Average speed = total distance ÷ time
taken = 356,000,000 km ÷ 172,800 s
$\approx 2,060$ km/s
$= 2,060 \times 3,600$ km/h
$= 7,416,000$ km/h

2. Speed of light ≈ 300 million meters
$\qquad\qquad = 300,000$ km/s
Santa's sleigh is moving at an average
speed of 2,060 km/s
$300,000 \div 2,060 = 145.631 \approx 145$
So, Santa is traveling at a speed 1/145
that of light.
Speed of sound ≈ 1,200 km/h = 1,200
÷ 3,600 km/s = 1/3 km/s
$2,060 \div (1/3) = 6,180$
So, Santa is traveling at a speed 6,180
times the speed of sound.
Note that when an object exceeds the
speed of sound, there will be at least
one sonic boom.

3. (a) 5,040 possible ways.
(b) With 25 cities, there are $1 \times 2 \times 3$
$\times \ldots \times 25$ possible ways.
In one year there are $3,600 \times 24$
$\times 365$ seconds.
Time taken = $(1 \times 2 \times 3 \times \cdots \times 25) \div$
$(3,600 \times 24 \times 365) \times (1,000,000)$
years
$\approx 4.9 \times 10^{11}$ years
$= 4.9 \times 10^2 \times 10^9$ years
$= 490 \times 10^9$ years
$= 490$ billion years

4. Christ was not born on December 25; there was no year zero.

Christ was probably born between 4 and 7 B.C., and not on 25 December.

6. 91.8 million homes.
$\dfrac{0.85 \times 378}{3.5} = 91.8$

25 Quickies and Trickies (p. 64)

1. The number of arrangements is the number of permutations of 25 elements:
$25! = 25 \times 24 \times 23 \times \cdots \times 3 \times 2 \times 1$

2. 169.
$1 = 1^2$
$1 + 3 = 2^2$
$1 + 3 + 5 = 3^2$
$\qquad \vdots$
$1 + 3 + 5 + \cdots + 25 = 13^2 = 169$

3. 8881784197.
$2.5^{25} = \left(\dfrac{25}{10}\right)^{25}$
$\qquad = \dfrac{25^{25}}{10^{25}}$
$\qquad = 8.881784197 \times 10^{34-25}$
$\qquad = 8.881784197 \times 10^9$
$\qquad = 8881784197$

4. $\underbrace{999\ldots99996}_{24\ 9s}$

5. A father calling daughter or son, or an aunt or uncle calling niece or nephew.

6. 13.
Several methods of solution exist. One method is as follows:

$S = 1 - 2 + 3 - 4 + + 25.$
$\ = (1 - 2) + (3 - 4) + \cdots + (23 - 24) + 25$
$\ = (-1) + (-1) + \cdots + (-) + 25$
$\ = (-1) \times 12 + 25$
$\ = -12 + 25$
$\ = 13$

7. One possible group of 25 nations is as follows:
Argentina, Brazil, Chile, Denmark, Egypt, Finland, Germany, Holland, India, Japan, Kenya, Laos, Malaysia, Norway, Oman, Peru, Qatar, Russia, Spain, Turkey, United States, Vietnam, Western Sahara, Yemen, Zambia—there is no country that begins with X.

8. The letter W is left out because it can always be written as UU—double U!

9. Each element in the 25-element set can either be included or not included in a subset. Hence, there are
$$\underbrace{2 \times 2 \times 2 \times \cdots \times 2}_{25} = 2^{25} \text{ subsets.}$$

10. Dodecagon.

11. $\dfrac{24}{25}$.

By partial fractions,
$$\frac{1}{n(n+1)} = \frac{1}{n} - \frac{1}{n+1}$$
$$S = \frac{1}{1 \times 2} + \frac{1}{2 \times 3} + \frac{1}{3 \times 4} + \ldots + \frac{1}{24 \times 25}$$
$$= \frac{1}{1} - \frac{1}{2} + \frac{1}{2} - \frac{1}{3} + \frac{1}{3} - \frac{1}{4} + \cdots + \frac{1}{24}$$
$$- \frac{1}{25}$$
$$= 1 - \frac{1}{25} = \frac{24}{25}$$

12. The word CHRISTMASTIME has 2 I's, 2 T's and 2 M's.
Hence the number of ways of arranging the letters is $\dfrac{13!}{2!2!2!}$
778,377,600

13. $2^{26} - 2$.
$$\begin{aligned} S &= 2 + 2^2 + 2^3 + \ldots + 2^{25} \\ 2S &= \qquad 2^2 + 2^3 + \cdots + 2^{25} + 2^{26} \end{aligned}$$
$$2S - S = -2 + \qquad\qquad\qquad 2^{26}$$
$$S = 2^{26} - 2$$
Hence, $2 + 2^1 + 2^2 + 2^3 + \cdots + 2^{25}$
$$= 2^{26} - 2$$

14. Rotations and reflections at a line through the center conserve neighboring relationships. Thus we have $\dfrac{25!}{2 \times 25} = \dfrac{24!}{2}$ distinct arrangements for more than 2 persons.

15. Numbers $25! + 2$, $25! + 3$, ..., $25! + 26$ will do the trick.

16. $2^{25} = (2^{24} - 1) + (2^{24} + 1)$.
A general result will be: If n is an integer greater than 1, then 2^n can be expressed as the sum of two consecutive integers. This follows from the relation $2^n = (2k - 1) + (2k + 1)$, which implies $k = 2^{n-2}$, thus resulting in $2^n = (2^{n-1} - 1) + (2^{n-1} + 1)$.
Use a similar argument to express 3^{25} as the sum of three consecutive integers.
$$3^{25} = (3^{24} - 1) + 3^{24} + (3^{24} + 1)$$

17. $\dfrac{12}{25}$
$$\frac{1}{1 \times 3} = \frac{1}{3} = \frac{1}{2} - \frac{1}{6}$$
$$\frac{1}{3 \times 5} = \frac{1}{15} = \frac{1}{6} - \frac{1}{10}$$
$$\frac{1}{5 \times 7} = \frac{1}{35} = \frac{1}{10} - \frac{1}{14}$$
$$\frac{1}{23 \times 25} = \frac{1}{46} - \frac{1}{50}$$
$$S = \frac{1}{1 \times 3} + \frac{1}{3 \times 5} + \frac{1}{5 \times 7} + \cdots + \frac{1}{23 \times 25}$$
$$= \frac{1}{2} - \frac{1}{6} + \frac{1}{6} - \frac{1}{10} + \frac{1}{10} - \frac{1}{14} + \cdots +$$
$$\frac{1}{46} - \frac{1}{50}$$
$$= \frac{1}{2} - \frac{1}{50}$$
$$= \frac{12}{25}$$

18. (a) One pair is a 2×2 square and a 4×1 rectangle.
Area of 2×2 square = 4 square units
Area of 4×1 rectangle = 4 square units
Perimeter of 2×2 square = 4×2 = 8 units
Perimeter of 4×1 rectangle = $2 \times (4 + 1)$ = 10 units

(b) Five such pairs exist:

2×1 and $\sqrt{2} \times \sqrt{2}$

2×2 and 4×1

5×1 and $\sqrt{5} \times \sqrt{5}$

4×2 and $\sqrt{8} \times \sqrt{8}$

5×2 and $\sqrt{10} \times \sqrt{10}$

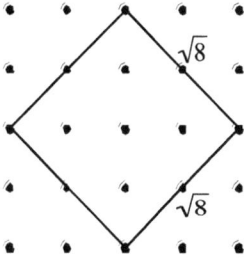

19. Note that $72^{2n+2} - 47^{2n} + 28^{2n-1} \equiv$
$(-3)^{2n+2} - (-3)^{2n} + 3^{2n-1} \equiv 3^{2n-1}(3^3 - 3 + 1) \pmod{25} = 0 \pmod{25}$

Note that
$(-3)^{2n+2} = 3^{2n+2}; (-3)^{2n} = 3^{2n}$

20. 600.

Observe that $\sqrt{x + 25} = 25$
$$x + 25 = 25^2$$
$$x = 25^2 - 25 = 600$$

21. 225 rectangles.

For a 2×2 board there are 9 rectangles:

4 1×1 squares
1 2×2 square
2 1×2 rectangles
2 2×1 rectangles
_____
9 rectangles

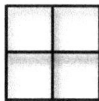

For a 3×3 board there are 36 rectangles.

For a 4×4 board there are 100 rectangles.
From the sequence $(1, 9, 36, ...)$, the number of rectangles that can be found on a 5×5 board is
$1^3 + 2^3 + \cdots + 5^3 = (1 + 2 + 3 + 4 + 5)^2$
$= 15^2 = 225$

22. $\dfrac{2^{26} - 1}{3 \times 2^{25}}$.

$S = 1 - \dfrac{1}{2} + \dfrac{1}{2^2} - \dfrac{1}{2^3} + \cdots - \dfrac{1}{2^{25}}.$

$= \left(1 + \dfrac{1}{2^2} + \dfrac{1}{2^2} + \cdots + \dfrac{1}{2^{24}}\right) - \left(\dfrac{1}{2} + \dfrac{1}{2^3} + \dfrac{1}{2^5} + \cdots + \dfrac{1}{2^{25}}\right)$

$\dfrac{1\left[1 - \left(\frac{1}{4}\right)^{13}\right]}{1 - \frac{1}{4}} - \dfrac{\frac{1}{2}\left[1 - \left(\frac{1}{4}\right)^{13}\right]}{1 - \frac{1}{4}}$

$= \dfrac{4}{3}\left[1 - \left(\dfrac{1}{2^2}\right)^{13}\right] - \dfrac{2}{3}\left[1 - \left(\dfrac{1}{2^2}\right)^{13}\right]$

$= \dfrac{2}{3}\left(1 - \dfrac{1}{2^{26}}\right)$

$= \dfrac{2}{3}\left(\dfrac{2^{26} - 1}{2^{26}}\right)$

$= \dfrac{2^{26} - 1}{3 \times 2^{25}}$

Alternatively,

$1 - \dfrac{1}{2} = \dfrac{1}{2} = \dfrac{1}{2^1}$

$\dfrac{1}{2^2} - \dfrac{1}{2^3} = \dfrac{1}{8} = \dfrac{1}{2^3}$

$\dfrac{1}{2^4} - \dfrac{1}{2^5} = \dfrac{1}{32} = \dfrac{1}{2^5}$

$\dfrac{1}{2^{24}} - \dfrac{1}{2^{25}} = \dfrac{1}{2^{25}}$

$S = \dfrac{1}{2} + \dfrac{1}{2^3} + \dfrac{1}{2^5} + \cdots + \dfrac{1}{2^{25}}$

$\dfrac{1}{2^2}S = \quad\quad \dfrac{1}{2^3} + \dfrac{1}{2^5} + \cdots + \dfrac{1}{2^{25}} + \dfrac{1}{2^{27}}$

$S - \dfrac{1}{2^2}S = \dfrac{1}{2} - \dfrac{1}{2^{27}}$

$\dfrac{3}{4}S = \dfrac{2^{26} - 1}{2^{27}}$

$S = \dfrac{4}{3}\left(\dfrac{2^{26} - 1}{2^{27}}\right) = \dfrac{2^{26} - 1}{3 \times 2^{25}}$

23. 25.

Let $x = \sqrt{25\sqrt{25\sqrt{25\sqrt{25}...}}}$
Then $x = \sqrt{25x}$
$$x^2 = 25x$$
$$x^2 - 25x = 0$$
$$x(x - 25) = 0$$
Since $x \neq 0$, $x = 25$
$\sqrt{25\sqrt{25\sqrt{25\sqrt{25}...}}} = 25$

24. $\dfrac{5+\sqrt{33}}{2}$.

Let $x = \sqrt{2 + 5\sqrt{2 + 5\sqrt{2 + 5\sqrt{2}}}...}$

Then $x = \sqrt{2 + 5x}$

$\qquad x^2 = 2 + 5x$

$\qquad x^2 - 5x - 2 = 0$

$x = \dfrac{-(-5) + \sqrt{(-5)^2 - 4(1)(-2)}}{2}$

$\quad = \dfrac{5 + \sqrt{33}}{2}$

Let $y = 5 + \dfrac{2}{5 + \dfrac{2}{5 + \dfrac{2}{5 + \cdots}}}$

$y = 5 + \dfrac{2}{y}$

$y^2 = 5y + 2$

$y^2 - 5y - 2 = 0$

$y = \dfrac{5 + \sqrt{33}}{2}$

25. 325.

Let $S = 1^2 - 2^2 + 3^2 - 4^2 + \cdots + 23^2 - 24^2$

$\begin{aligned}
1^2 - 2^2 &= -(2-1) \times (2+1) &= -3 \\
3^2 - 4^2 &= -(4-3) \times (4+3) &= -7 \\
5^2 - 6^2 &= -(6-5) \times (6+5) &= -11 \\
&\quad\vdots &\vdots \\
23^2 - 24^2 &= -(24-23) \times (24+23) &= -47
\end{aligned}$

$S = -3 \ + -7 \ + \cdots + -47$

$S = -47 \ + \ -43 + \cdots + \ -3$

$2S = -(3 + 7 + 11 + \cdots + 47)$

$\quad = -50 \times 12 = -600$

$S = -300$

Hence, $1^2 - 2^2 + 3^2 - 4^2 + \cdots + 25^2$

$\qquad = S + 25^2$

$\qquad = -300 + 625$

$\qquad = 325$

A Formula for Christmas Day (p. 73)

1. (i) Saturday.
(ii) Thursday.
(iii) Thursday.
(iv) Monday.

(i) 2004
C = 20, Y = 04
K = 20/4 = 5; G = 04/4 = 1
D = 50 + 4 + 5 + 1 − 40 = 20
20/7 = 2 R 6
So, 25 December 2004 falls on a Saturday.

(ii) 1986
C = 19, Y = 86
K = 19/4 = 4.75, rounded off to 4.
G = 86/4 = 21.5, rounded off to 21.
D = 50 + 86 + 4 + 21 − 38 = 123
123/7 = 17 R 4
So, 25 December 1986 fell on a Thursday.

(iii) 2025
C = 20, Y = 25
K = 20/4 = 5
G = 25/4 = 6.25, rounded off to 6.
D = 50 + 25 + 5 + 6 − 40 = 46
46/7 = 6 R 4
So, 25 December 2025 falls on a Thursday.

(iv) 2215
C = 22, Y = 15
K = 22/4 = 5.5, rounded off to 5.
G = 15/4 = 3.75, rounded to 3.
D = 50 + 15 + 5 + 3 − 44 = 29
29/7 = 4 R 1
So, 25 December 2215 falls on a Monday.

Number of zeros in 1 × 2 × 3 × ··· × 25 (p. 85)

1. (a) 6   (b) 9
(c) 1   (d) 1   (e) 3

2. In $1 \times 2 \times 3 \times \cdots \times 16 \times 17$, there are three terms with 5 as a factor: 5, 10, 15. The product of these three 5's and three even numbers will yield three terminal zeros.

3. Seven 5's.
The numbers which have 5 as a factor are: 5, 10, 15, 20, 25, 30.
These numbers will give seven 5's.
(Remember: 25 = 5 × 5)

4. 4 zeros.

Consider $1 \times 2 \times 3 \times \cdots \times 19 \times 20$.
Terms with 5 as a factor are: 5, 10, 15, 20.
Hence, we have a total of four 5's, thus yielding 4 zeros.

5. 24 zeros.

In $1 \times 2 \times 3 \times \cdots \times 99 \times 100$, the number of terms with 5's are all the multiples of 5:
5, 10, 15, 20, 25, 30, 35, 40, 45, 50, 55, 60, 65, 70, 75, 80, 85, 90, 95, 100
Out of the 20 of them, four have two 5's:
25 (5×5), 50 $(2 \times 5 \times 5)$, 75 $(3 \times 5 \times 5)$ and 100 $(4 \times 5 \times 5)$.
There are altogether $(20 + 4) = 24$ of them.
Hence there are 24 zeros at the end of $1 \times 2 \times 3 \times \cdots \times 99 \times 100$.

6. 12 zeros.

$2 \times 4 \times 6 \times 8 \times 10 \times 12 \times \cdots \times 100$
$= 10 \times 20 \times 30 \times 40 \times 50 \times 60 \times 70 \times 80 \times 90 \times 100 \times \cdots$
Each even number (having a factor 5) ending with zero contributes one zero.
Each number having a factor (5×5) contributes two zeros—there are two such numbers: 50 $(2 \times 5 \times 5)$ and 100 $(4 \times 5 \times 5)$.
Number of ending zeros $= 10 + 2 = 12$

7. 124 zeros.

$1000 \times 998 \times 996 \times \cdots \times 6 \times 4 \times 2$
$= 2 \times 4 \times 6 \times 8 \times \cdots \times 998 \times 1000$
$= 10 \times 20 \times 30 \times 40 \times \cdots \times 990 \times 1000 \times \cdots$
Each of the numbers (10, 20, ..., 1000) contributes one zero—there are 100 such numbers.
Each of the numbers (50, 100, 150, 200, 250, ..., 900, 950, 1000) contributes two zeros—there are 20 such numbers.
Each of the numbers (250, 500, 750, 1000) contributes three zeros—there are 4 such numbers.
Number of ending zeros $= 100 + 20 + 4 = 124$

8. 47.

The number of terms divisible
by 2 is 25,
by 4 is 12,
by 8 is 6,
by 16 is 3, and
by 32 is 1.
The maximum number of divisions by 2 is $25 + 12 + 6 + 3 + 1 = 47$.

9. If $1 \times 2 \times 3 \times \ldots \times N$ does not end in a zero, that means that 10 does not divide it. In other words, not both 2 and 5 divide the product. This leaves only five possible cases: 0, 1, 2, 3, and 4.

10. 18 zeros.

$\frac{100!}{25!} = 26 \times 27 \times \cdots \times 99 \times 100$

In $26 \times 27 \times 28 \times \cdots \times 99 \times 100$, the number of terms with 5's are all the multiples of 5:
30, 35, 40, 45, 50, 55, 60, 65, 70, 75, 80, 85, 90, 95, 100
Out of the 15 of them, three have two 5's:
50 $(2 \times 5 \times 5)$, 75 $(3 \times 5 \times 5)$, and 100 $(4 \times 5 \times 5)$.
There are altogether $(15 + 3) = 18$ of them.
Hence there are 18 zeros at the end of $\frac{100!}{25!}$.

11. 249 zeros.

The number of 5's in 1000! is $200 + 40 + 8 + 1 = 249$.
There are enough of 2's to match each 5 to get a 10.
Thus, 1000! ends in 249 zeros.

12. $5^{10}!$ ends in 2,441,406 zeros.
In $5^{10}!$, $5^{10}/5 = 5^9$ factors are divisible by 5, $5^{10}/5^2 = 5^8$ are divisible by 5 twice, $5^{10}/5^3 = 5^7$ are divisible by 5 three times, and so on, for a total T of
$T = 5^9 + 5^8 + 5^7 + \cdots + 5^1 + 1.$

To compute T easily, note that
$5T = 5^{10} + 5^9 + 5^8 + \cdots + 5^1$.
When you subtract T from $5T$, most terms cancel, and you get
$4T = 5^{10} - 1$
and
$T = (5^{10} - 1) \div 4 = 2{,}441{,}406$

In general, $5^N!$ has $\dfrac{5^N - 1}{4}$ terminal zeros.

$1 \times 2 \times 3 \times \ldots \times (n - 1) \times n$ ends in 25 zeros (p. 92)

1. (a) $n = 95, 96, 97, 98,$ and 99
   (b) Since multiplying $1 \times 2 \times \cdots \times 99$ by 100 adds two more zeros, so there is no number n for which $1 \times 2 \times 3 \times \cdots \times n$ has exactly 23 zeros at the end.

2. Earlier we saw that the product
   $1 \times 2 \times \cdots \times 25$ has 24 zeros.
   $\underbrace{5, 10, 15, \ldots, 90, 95, 100,}_{24\ 0\text{'s}}$

   $\underbrace{105 - 120,}_{4\ 0\text{'s}} \underbrace{125,}_{3\ 0\text{'s}} \underbrace{130 - 145,}_{4\ 0\text{'s}} \underbrace{150}_{2\ 0\text{'s}}$
   $(105, 110, 115, 120)$
   Total number of zeros $= 24 + 4 + 3 + 4 + 2 = 37$

   Note that the next multiple of 5 (i.e., 155) will raise the total to 38 zeros. So, the values of n for which $1 \times 2 \times 3 \times \cdots \times n$ ends in 37 zeros are 150, 151, 152, 153, and 154.

3. $n = 110, 111, 112, 113, 114.$

4. $n = 165, 166, 167, 168,$ and $169.$
   The integer n will end in exactly 40 zeros if and only if the largest power of 5 that divides $n!$ is 5^{40}.
   If $n = 125$, then $M_n = \lfloor 125/5 \rfloor + \lfloor 125/25 \rfloor + \lfloor 125/125 \rfloor = 25 + 5 + 1 = 31$

If $n = 200$, then $M_n = \lfloor 200/5 \rfloor + \lfloor 200/25 \rfloor + \lfloor 200/125 \rfloor = 40 + 8 + 1 = 49$
So if $M_n = 40$, then $125 < n < 200$.

Let $n = 125 + 25s + 5u + v$, where $s = 0, 1$ or 2, or $0 \le u \le 4$, and $0 \le v \le 4$.
Then $M_n = 1 + (5 + s) + (25 + 5s + u) = 31 + 6s + u$, thus $M_n = 40$ iff $6s + u = 9$.

Since $u \le 4$, we must have $s = 1$ and $u = 3$.
$\therefore n = 165 + v$ for $v = 0, 1, \ldots, 4$
Thus $n!$ ends in exactly 40 zeros for $n = 165, 166, 167, 168,$ and $169.$

5. From question 2, $1 \times 2 \times 3 \times \cdots \times 145$ has 35 zeros.
   But the next multiple of 5 (i.e., 150) contributes 2 zeros. So the possible values of n for which $n!$ has exactly 35 zeros are $n = 145, 146, 147, 148,$ and 149.

6. It suffices to consider the largest power of 5 that divides $n!$.
   For $n = 1000$, $\lfloor 1000/5 \rfloor + \lfloor 1000/25 \rfloor + \lfloor 1000/125 \rfloor + \lfloor 1000/625 \rfloor = 249$, but the corresponding sum for $n = 999$ is 246. Thus $n!$ cannot end in either 247 or 248 zeros.

7. Let $n! = 5^s \times k$ where k is not divisible by 5, i.e., s is the number of factor 5 in $n!$.
   Each such factor 5 multiplied by a factor 2 will contribute a zero in $n!$. Since there are more factors of 2 than those of 5 in $n!$, thus the value of s will give us the exact number of terminal zeros in $n!$. Thus
   $s = \lfloor n/5 \rfloor + \lfloor n/25 \rfloor + \lfloor n/125 \rfloor + \cdots \le$

   $n\left(\dfrac{1}{5} + \dfrac{1}{25} + \dfrac{1}{125}\right) = \dfrac{n}{4}$
   So, if $s = 2003$, $n \le 2003 \times 4 = 8012$.

   When $n = 8012$, $s = 1602 + 320 + 64 + 12 + 2 = 2000$. We need at least 3 more factors of 5.

Since 8015 and 8020 each contribute one more factor 5, the smallest possible n must be 8025.

9. The highest power of 5 that divides 29! is $\lfloor 29/5 \rfloor + \lfloor 29/5^2 \rfloor = 5 + 1 = 6$.

The highest power of 2 that divides 29! is
$\lfloor 29/2 \rfloor + \lfloor 29/2^2 \rfloor + \lfloor 29/2^3 \rfloor + \lfloor 29/2^4 \rfloor$
$= 14 + 7 + 3 + 1 = 25$

Thus the highest power of 4 that divides N is 9.

The highest power of 3 that divides 29! and hence N is
$\lfloor 29/3 \rfloor + \lfloor 29/3^2 \rfloor + \lfloor 29/3^3 \rfloor = 9 + 3 + 1$
$= 13$.
So the highest power of 12 that divides N is 9.

Taking up the Cross (p. 94)

1. 22 squares (= $5 \times 2 + 3 \times 2 + 3 \times 2$).

2.

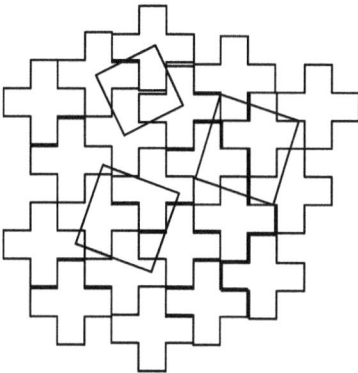

The Greek cross tessellates, yielding an infinite number of dissections of the cross into a square.

Take any four corresponding points of the tessellation, and a dissection of the cross into a square is obtained.

This simplicity occurs as the cross is made up of five unit squares, with five being the sum of two squares:
$$5 = 2^2 + 1^2.$$

This condition is not sufficient. All the other pentominoes (shapes formed by laying five identical squares against each other, complete edge to complete edge) satisfy the condition, but only some of them will tessellate, leading to similar dissection.

3.

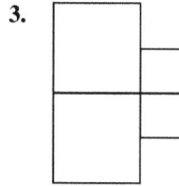

4. 30 squares = 35 squares − 5 squares

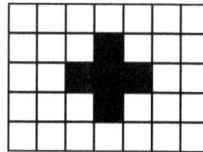

6.

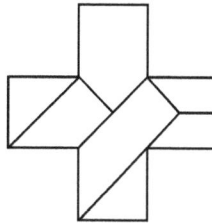

7.

8.

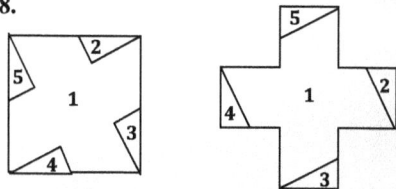

9.

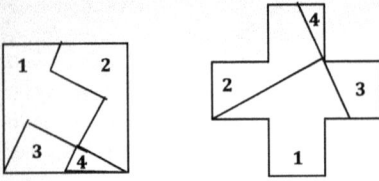

10.

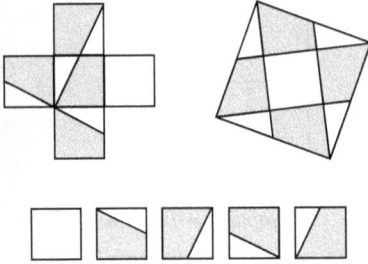

11.

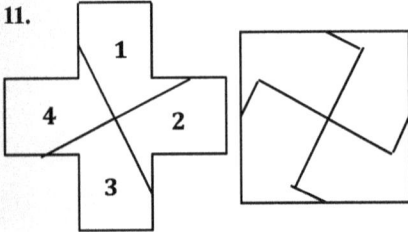

12.

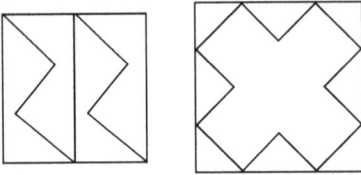

13.

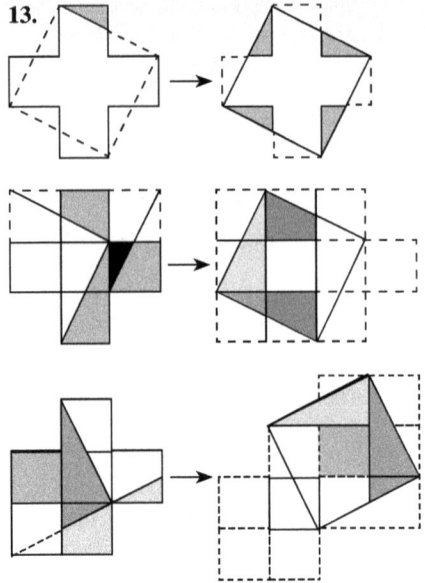

14.

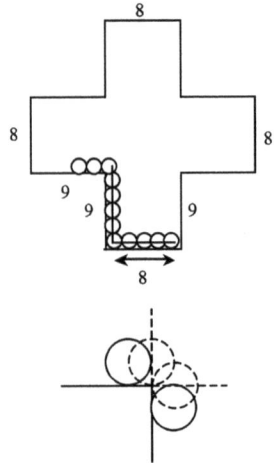

Total distance traveled
$= 9 \times 8 + 8 \times 4 + 2\pi$
$= (104 + 2\pi)$ cm

15.

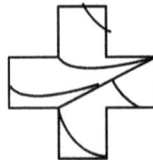

Number of Digits in 25²⁵ (p. 106)

1. 10^{25}.

$$2^{83} = 2^3 \cdot 2^{80}$$
$$= 8 \times 2^{10 \times 8}$$
$$= 8 \times (2^{10})^8$$
$$\approx 8 \times (10^3)^8$$
$$= 8 \times 10^{24}$$
$$\approx 10^{25}$$

2. 2^{125}.

$$2^{125} = 2^5 \times 2^{120}$$
$$= 32 \times (2^{10})^{12}$$
$$> 32 \times (10^3)^{12}$$
$$= 32 \times 10^{36}$$

Thus, 2^{125} is larger than 32×10^{36}.

3. $n = 11$.

$$24 \le \log n^{24} < 25$$
$$24 \le 24 \log n < 25$$
$$\frac{24}{24} \le \log n < \frac{25}{24}$$
$$10^1 \le n < 10^{\frac{25}{24}}$$
$$10 \le n < 11.01$$

4. $n = 72$.

Let $25^n > 10^{100}$

Taking log on both sides, we have
$$\log 25^n > \log 10^{100}$$
$$n \log 25 > 100$$
$$n > \frac{100}{\log 25} \approx 71.53$$

The smallest integer that is larger than 71.53 is 72.

Hence we take n to be 72.

5. $n = 10^{101}$

$$2524 \le \log n^{25} < 2525$$
$$2524 \le 25 \log n < 2525$$
$$\frac{2524}{25} \le \log n < \frac{2525}{25}$$
$$100.96 \le \log n < 101$$
$$10^{100.96} \le n < 10^{101}$$

So n can only be 10^{101}.

Check: $\log (10^{101})^{25} = 25 \times 101 \log 10$
$$= 2525$$

6. Since 5^{10} is close to 10^7, we have $5 \approx 10^{0.7}$; so $5^{55} \approx 10^{0.7 \times 55} = 10^{38.5}$

Thus, 5^{55} has roughly 39 digits.

Now, $5^{555} \approx 10^{0.7 \times 555} = 10^{388.5}$, so it has roughly 389 digits (in fact, the answer is 388).

Finally, $5^{5^{55}}$ has roughly 1.94×10^{38} digits.

Christmas Tangrams (p. 109)

Festive shapes

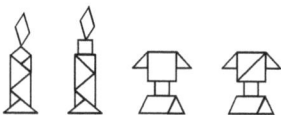

1.

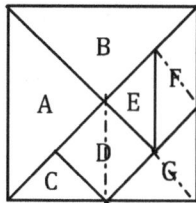

Area of piece A = Area of piece B
$$= \frac{1}{4} \text{ cm}^2$$

Area of piece C = Area of piece E
$$= \frac{1}{16} \text{ cm}^2$$

Area of piece D = Area of piece F
= Area of piece G = $\frac{1}{4}$ cm²

2.

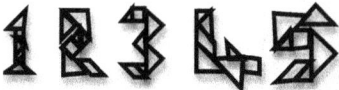

3.

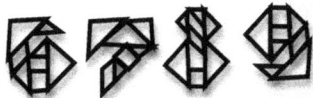
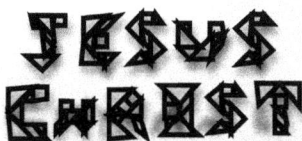

4.

Notice that in both cases the head, hat, and arm are exactly alike, and the width at the base of the body is the same. In the first case, the body is made up of four pieces, and in the second design only three.

The first is larger than the second by exactly that narrow strip indicated by the dotted line between A and B. This strip is therefore exactly equal in area to the piece forming the foot in the other design, although when distributed along the side of the body the increased dimension is not easily apparent to the eye.

The above paradox is what we call the principle of concealed distribution.

5. (a)   (b)

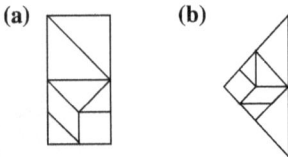

(c)

6.

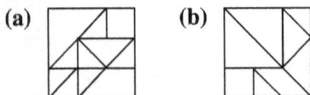

7. (a)   (b)

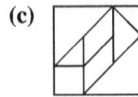

(c)

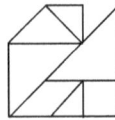

9. (a)   (b)

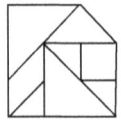

10.

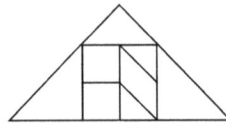

What Day Is Christmas 2025? (p. 124)

1. (a) Tuesday
    (b) Thursday
    (c) Saturday

2. The calendar in any year is determined by two factors:
- whether the year is a leap year or not;
- the day of the week on January 1.

From a non-leap year to the following year, the day of the week of January 1 advances by one day in the list of 7 days of the week (Monday, Tuesday, …, Sunday). On the other hand, from a leap year to the following year, the day of the week of January 1 advances by two days, since the leap year has an extra day (February 29).

Let's tabulate the day of the week on which January 1 falls in a given year, from 2008 onwards.

From the table, notice that although January 1 falls on a Tuesday in 2013, 2019, and 2030, they are all non-leap years. The next leap year when the 2008 calendar could be re-used entirely would be in the year 2036—28 years later.

Year	Day of the week on January 1
2008 (leap)	Tuesday
2009	Thursday
2010	Friday
2011	Saturday
012 (leap)	Sunday
2013	*Tuesday*
2014	Wednesday
2015	Thursday
2016 (leap)	Friday
2017	Sunday
2018	Monday
2019	*Tuesday*
2020 (leap)	Wednesday
2021	Friday
2022	Saturday
2023	Sunday
2024 (leap)	Monday
2025	Wednesday
2026	Thursday
2027	Friday
2028 (leap)	Saturday
2029	Monday
2030	*Tuesday*
2031	Wednesday
2032 (leap)	Thursday
2033	Saturday
2034	Sunday
2035	Monday
2036 (leap)	**Tuesday**
2037	Thursday
2038	Friday

3. January 1 in 2010 falls on a Friday. From the table, January 1 falls on a Friday in 2016, but it is a leap year. So the 2010 calendar could not be reused.

January 1 again falls on a Friday in 2021, 2027, and 2038, all of which are non-leap years. So the 2010 calendar could be reused in these future years.

The Answer Is Not 25
(p. 130)

1. (a) The nth term is
$$\frac{1}{6}(n-1)(n-2)(n-3)+n.$$

(b) $x = 40$, $d = 25$, $p = \dfrac{40-25}{24}$
$$= \frac{15}{24} = \frac{5}{8}$$
The nth term is $\dfrac{5}{8}(n-1)(n-2)$
$(n-3)(n-4)+n^2$.

Check: $n = 5$, $\dfrac{5}{8} \times 4 \times 3 \times 2 \times 1$
$+ 25 = 15 + 25 = 40$

(c) The nth term is $\dfrac{15}{24}(n-1)(n-2)$
$(n-3)(24-5n)+n$.
For the sequence 1, 2, 3, 5, the
formula is $\dfrac{1}{6}(n-1)(n-2)(n-3)+n$
[From (a)]
The above formula or expression gives
$$1, 2, 3, 5, 9$$
when 1, 2, 3, 4, 5 is substituted for n
in order. [$n = 5$, $\dfrac{1}{6} \times 4 \times 3 \times 2 + 5 = 9$]
Now, when $n = 5$, $(n-1)(n-2)$
$(n-3)(n-4) = 4 \times 3 \times 2 \times 1 = 24.$
But since we want the 5th term to be
4 instead of 9, we have
$d = 9$, $x = 4$; so $p = \dfrac{x-d}{24} = \dfrac{4-9}{24}$
$= -\dfrac{5}{24}$
Therefore, the formula that will yield
$$1, 2, 3, 5, 4$$
is $-\dfrac{5}{24}(n-1)(n-2)(n-3)(n-4)+$
$\left[\dfrac{1}{6}(n-1)(n-2)(n-3)+n\right]$, which
simplifies to
$$\frac{1}{24}(n-1)(n-2)(n-3)(24-5n)+n.$$

Check: $n = 4$, the value of the
expression is $\dfrac{1}{24} \times 3 \times 2 \times 1 \times 4 + 4$
$= 1 + 4 = 5$
$n = 5$, the value of the expression is
$\dfrac{1}{24} \times 4 \times 3 \times 2 \times -1 + 5 = -1 + 5 = 4$

A Christmas Potpourri
(p. 133)

1. 4 times.

2. 300 handshakes.
    $24 + 23 + 22 + \cdots + 3 + 2 + 1 = 300$

3. The letters C and R do not belong. All the others have rotational symmetry.

4. 0 volunteers.

    6 persons can deliver 6 gifts in 6 minutes.
    6 persons can deliver 25 gifts in 25 minutes.
    So no extra volunteers are needed.

5. 325.

    $$\begin{array}{r} 1 + 2 + 3 + \cdots + 24 + 25 \\ 25 + 24 + 23 + \cdots + 2 + 1 \\ \hline 26 + 26 + 26 + \cdots + 26 + 26 \end{array}$$

    $\text{Sum} = \dfrac{26 \times 25}{2} = 13 \times 25$
    $= \dfrac{1300}{4}$
    $= 325$

    Alternatively,
    $\text{Sum} = 1 + 27 \times \dfrac{24}{2} = 325$

6. 107.5°.

7. $5 \div .2 = 25$

8. (a) At least 60 triangles,
            10 diamonds,
            10 kites,
            7 pentagons
    (b) Trapezoids, stars, quadrilaterals (convex and concave), and so on.

9.
Year	Number of stems	Number of cones
1	$1 = 2^0$	$2 = 2^1$
2	$2 = 2^1$	$4 = 2^2$
3	$4 = 2^2$	$8 = 2^3$
4	$8 = 2^3$	$16 = 2^4$
...	...	...
25	2^{24}	2^{25}

Number of stems $= 1 + 2^1 + 2^2 + \cdots +$
$2^{24} = \dfrac{1 \times (2^{25} - 1)}{2 - 1} = 2^{25} - 1$
Number of cones $= 2^1 + 2^2 + 2^3 + \cdots +$
$2^{25} = \dfrac{2 \times (2^{25} - 1)}{2 - 1} = 2(2^{25} - 1)$

10. 364 days ($= t_1 + t_2 + \cdots + t_{12}$).

11. 364/1.

    The baby can be born on any date of the year. Since there are 365 dates in the year, of which December 25 is one, the odds that the baby—unless induced—would be born on that date are 364 to 1.

12. $25 = 4! + (\sqrt{4} + \sqrt{4})/4$.

13. SEASON GREETINGS (C's on greetings.)

14. The digit 7.

Digit	Number of digits
1 – 9	9
10 – 99	$90 \times 2 = 180$
100 – 999	$900 \times 3 = 2700$

So the 2525th digit would come from the digit of a 3-digit number.

Number of digits from 3-digit numbers $= 2525 - (9 + 180) = 2336$

$2336 = 778 \times 3 + 2$

$100 - 877 \ 878$
           $\uparrow$
        2525th digit

15. 12 noon.

    25 days, 25 hours, 25 minutes
    = 26 days, 1 hour 25 minutes
    1:25 − 1 hr 25 mins = 12 noon

16. A father calling daughter or son, or an aunt or uncle calling niece or nephew.

17. Oct. 31 = Dec. 25

    If "Oct." is the short form for "octal" and "Dec." for "decimal," then 31 (in base-8 notation) is equal to 25 (in base-10 notation.)

$$31_8 = 25_{10}$$

Comments

Suzanne L. Hanauer:

Oct. 31 can be written 10/31 or 1031. Dec. 25 is 12/25 or 1225. And $1,031 = 1,225$ (modulo 194).

Let the five letters in Oct. and Dec. stand for digits as follows:

O = 6, C = 7, T = 5, D = 8, E = 3

We can decode Oct. 31 = Dec. 25 as:

$$675 \times 31 = 837 \times 25 = 20,925$$

18. Let $x = 0.039999...$

Then $100x = 3.9999...$

$$99x = 3.96000...$$
$$x = 3.96/99$$
$$= 396/9900$$
$$= 1/25$$

Hence, $0.039999... = 1/25$

19. 1.282 431 813

Consider the expansion of $(1 + x^2)^{25}$.

$$(1 + x^2)^{25} = 1 + \binom{25}{1}x^2 + \binom{25}{2}(x^2)^2 +$$
$$\binom{25}{3}(x^2)^3 + \binom{25}{4}(x^2)^4 + \binom{25}{5}(x^2)^5 + \cdots$$
$$= 1 + 25x^2 + 300x^4 + 2300x^6$$
$$+ 12,650x^8 + 53,130x^{10} + \cdots$$

Comparing $(1.01)^{25}$ with $(1 + x^2)^{25}$, we have $1 + x^2 = 1.01$, which gives $x = \pm 0.1$

$$(1.01)^{25} = [1 + (0.1)^2]^{25}$$
$$= 1 + 25(0.1)^2 + 300(0.1)^4 +$$
$$2300(0.1)^6 + 12,650(0.1)^8$$
$$+ 53,130(0.1)^{10} + \cdots$$
$$= 1 + 0.25 + 0.03 + 0.0023$$
$$+ 0.000\ 126\ 5 + 0.000\ 005$$
$$313 + \cdots$$
$$\approx 1.282\ 431\ 813$$

20. Let A be the first guard and B the second.

If A and B are in the same row, A is taller. Similarly, if A and B are in the same column, B is shorter.

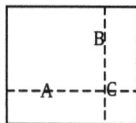

If they are in different columns and rows, let C be the guard in A's row and B's column. C is shorter than A, but taller than B. So A is always taller than B.

21. Rotations and reflections at a line through the center conserve neighboring relationships. Thus there are $\dfrac{25!}{2 \times 25} = \dfrac{24!}{2}$ distinct arrangements for more than 2 persons.

22. Let a represents the initial premium, and r the annual increase.

Then $a = 25\ 000$ and $r = 1.025$

(i) Amount earned in her 25th year salary
$$= ar^{n-1}$$
$$= \$25,000 \times 1.025^{24}$$
$$\approx \$45,218.15$$

(ii) $S_{25} = \dfrac{a(1 - r^n)}{1 - r}$
$$= \dfrac{25,000 \times (1 - 1.025^{25})}{1 - 1.025}$$
$$\approx 853,949.10$$

So she will earn a total of $853,949.10.

(iii) $S_n = \dfrac{25,000\ (1 - 1.025^n)}{1 - 1.025}$
$$\geq 1,000,000$$
$$\dfrac{25}{0.025}(1.025^n - 1) \geq 1000$$
$$1.025^n - 1 \geq 1$$
$$1.025^n \geq 2$$
$$n \lg 1.025 \geq \lg 2$$
$$n \geq \dfrac{\lg 2}{\lg 1.025}$$
$$n \geq 28.08$$

So she has to work a minimum of 29 years before her total earnings exceeds $1,000,000.

23. $25 = 9\sqrt{9} - .\overline{9} - .\overline{9}$
$24 = 9 + 9 + 9 - \sqrt{9}$; $23 = 9\sqrt{9} - \sqrt{9} - .\overline{9}$

24. 1. $(9 - 9) + (9 - 9) = 0$
2. $(\sqrt{9} - \sqrt{9}) + (\sqrt{9} - \sqrt{9}) = 0$
3. $(9 - 9) - (9 - 9) = 0$

4. $(\sqrt{9} - \sqrt{9}) - (\sqrt{9} - \sqrt{9}) = 0$
5. $(9 + 9) - (9 + 9) = 0$
6. $(\sqrt{9} + \sqrt{9}) - (\sqrt{9} + \sqrt{9}) = 0$
7. $(9 - 9) - (\sqrt{9} - \sqrt{9}) = 0$
8. $(9 / 9) - (9 / 9) = 0$
9. $(\sqrt{9} / \sqrt{9}) - (9/9) = 0$
10. $(\sqrt{9} / \sqrt{9}) - (\sqrt{9} / \sqrt{9}) = 0$
11. $9 \times 9 - 9 \times 9 = 0$
12. $\sqrt{9} \times \sqrt{9} - \sqrt{9} \times \sqrt{9} = 0$
13. $(9 \sqrt{9} / 9) - \sqrt{9} = 0$
14. $9 - 9(9/9) = 0$
15. $9 - 9(\sqrt{9} / \sqrt{9}) = 0$
16. $9 \times \sqrt{9} - 9 \times \sqrt{9} = 0$
17. $9 - \sqrt{9} - \sqrt{9} - \sqrt{9} = 0$
18. $(9 \sqrt{9} / \sqrt{9}) - 9 = 0$
19. $(\overline{.9} + \overline{.9}) - (\overline{.9} + \overline{.9}) = 0$
20. $(\overline{.9} - \overline{.9}) - (\overline{.9} - \overline{.9}) = 0$
21. $(\overline{.9} - \overline{.9}) + (\overline{.9} - \overline{.9}) = 0$
22. $(\overline{.9}/\overline{.9}) - (\overline{.9}/\overline{.9}) = 0$
23. $(9 - 9) + (\overline{.9} - \overline{.9}) = 0$
24. $(\overline{.9} - \overline{.9}) + (\sqrt{9} - \sqrt{9}) = 0$
25. $(9 - 9) - (\overline{.9} - \overline{.9}) = 0$

25. 1.

Assume that n children went to the party. Let's label them 1, 2, ..., n. Let's introduce the random variables

$$X_i = \begin{cases} 1 & \text{if the } i\text{th child carries home} \\ & \text{his or her original present} \\ 0 & \text{otherwise} \end{cases}$$

The total number of children carrying home their original present is
$$S = X_1 + \cdots + X_n.$$
Presents can be ordered in $n!$ ways. There exist $(n - 1)!$ orderings such that the present brought by the ith child is in the ith place. Thus

$$P(X_i = 1) = \frac{(n-1)!}{n!} = \frac{1}{n}.$$

Moreover,
$$E(X_i) = 1 \times P(X_i = 1) + 0 \times P(X_i = 0)$$
$$= \frac{1}{n}$$
and
$$E(S) = E(X_1) + \cdots + E(X_n) = 1$$

The expected number of children carrying home their original presents is 1 independent of n.

CHRISTMAS *Alphametics*
(p. 138)

1. (a) $5\overline{)31{,}770}$, $6\overline{)18{,}990}$, $6\overline{)19{,}002}$, $8\overline{)35{,}112}$.

   (b) $4\overline{)14{,}992}$, $7\overline{)67{,}004}$.

2. (a) $2 + 97{,}445 + 6{,}928$.

   (b)
   $$\begin{array}{r} 24{,}794 \\ -\ 16{,}452 \\ \hline 8{,}342 \end{array} \qquad \begin{array}{r} 36{,}156 \\ -\ 28{,}693 \\ \hline 7{,}463 \end{array}$$

3. $139 + 139 + \cdots + 139 = 9730$.
   $139 \times 70 = 9730$.

4. $366 + 9780 = 1572 + 8574$.

5. $8180 + 939 + 7854 + 8296 = 25{,}869$, with 0 and 4 interchangeable.

6. The solution with the maximum total is on the right.

   The values for D and E (3 and 6) are interchangeable, so are V and O (2 and 5). There are 8 different solutions, or 10 if A is allowed to have the value zero.

   $$\begin{array}{r} 8384 \\ 803 \\ 626 \\ 50 \\ 8 \\ \hline 9871 \end{array}$$

7.
$$\begin{array}{r} 9387 \\ +\ 9387 \\ \hline 18{,}774 \end{array}$$

8. Hint: To obtain the simplest fraction equivalent to a decimal of n repeating digits, put the repeating period over n 9's and reduce to its lowest terms.

$$\frac{2421}{303} = .79867986\ldots$$

Let F = .TALKTALKTALK…
Then 1000F = TALK.TALKTALK…
Subtracting, 9999F = TALK

Then $\dfrac{EVE}{DID} = F = \dfrac{TALK}{9999}$

Therefore $\dfrac{TALK}{9999}$ when reduced to lowest terms equals $\dfrac{EVE}{DID}$.

Hence the denominator, "DID" is a 3-digit factor of 9999, namely 101, 303, or 909.

(a) Assume DID = 101.

Then $\dfrac{EVE}{101} = \dfrac{TALK}{9999}$

TALK = (EVE)99 = (EVE)(100 − 1)
= EVE00 − EVE

This leaves an E as the first digit which therefore cannot be T.
Therefore DID ≠ 101.

(b) Assume DID = 909.

Then $\dfrac{EVE}{909} = \dfrac{TALK}{9999}$

TALK = (EVE)11 = (EVE)(10 + 1) = EVE0 + EVE

The units digit is E and therefore cannot be K.
Therefore DID ≠ 909.

(c) Hence if there is a solution DID = 303.

Since F is a proper fraction, EVE can be only (1*1) or (2*2), leaving for trial only 121; 141; 151; 161; 171; 181; 191; also 212; 242; 252; 262; 272; 282; 292.

All except 242 repeat, in the quotient, a digit that appears in the dividend so that finally 242/303 = .79867986… is the only solution.

Alternatively,

$\dfrac{EVE}{DID}$ = .TALKTALKTALK…

$\dfrac{EVE}{DID} = \dfrac{TALK}{9999}$

Hence, DID is a factor of 9999. The only three factors that divide DID are 101, 303, and 909.

If DID = 101, then $\dfrac{EVE}{101} = \dfrac{TALK}{9999}$ and

EVE = $\dfrac{TALK}{99}$.

Rearranging terms,
TALK = 99 × EVE

EVE cannot be 101, since we have assumed 101 to be DID, and anything larger than 101, when multiplied by 99, has a 5-digit product. And so DID = 101 is rejected.

If DID = 999, then $\dfrac{EVE}{909} = \dfrac{TALK}{9999}$ and

EVE = $\dfrac{TALK}{11}$.

Rearranging terms,
TALK = 11 × EVE

In that case the last digit of TALK would have to be E. Since it is not W, 909 is also excluded.

Only 303 is left as a possibility for DID. Since EVE must be smaller than 303, E is 1 or 2.

Of the 14 possible values (121, 141, …, 292) only 242 produces a decimal fitting TALKTALK…, in which all the digits differ from those in EVE and DID.

The unique answer is

$\dfrac{EVE}{909}$ = .798679867986…

If $\dfrac{EVE}{DID}$ is not assumed to be in lowest terms, there is one other solution, $\dfrac{212}{606}$ = .349834983498…, proving, as Joseph Madachy has remarked, that EVE double-talked.

9. (tw001) (9t57) = 830e72t57

10.

9,486	9,476	9,386	9,376
+ 1,076	+ 1,086	+ 1,076	+ 1,086
10,562	10,562	10,462	10,462

11.

  29,661
  29,661
  29,661
+ 3,910
  92,893

12. P = 1, L = 0, E = 8 or 9.

If E = 8, then X + M + A = E + 19 so X + M + A = 27, which is impossible. Hence E = 9. Then, S + Y = 9.

Also, X + M + A = 18, so A > 2, M > 2.

Based on S + Y = 9, we tabulate the possible values.

S = 2			etc.
Y = 7			etc.

A + I → 2		
A = 8	4	
I = 4	8	

M + R = 9		9	
R = 3 6	3		6
M = 6 3	6		3
X = ~~4~~ ~~7~~	~~8~~		~~11~~

13. 9597 + 8 + 121 + 395 + 419 + 94 = 10,634.

14. √(4,133,089) = 2033.

First, *KISS* < √(10,000,000) = 3162..., so *K* is one of 1, 2, 3; and if *K* is 3 then $1 ≤ 1$. And *S* must be one of 2, 3, 4, 7, 9 since *PASSION* does not end in *S* (8 is eliminated because 88 × 88 = 7744). Because SS^2 and in *ON* and ISS^2 ends in *ION*, we get the following chart of possibilities.

S	ON	I	K	KISS	
2	84	-			
3	89	0	1 or 2	1033	2033
		2	1 or 2	1233	2233
		4	1 or 2	1433	2433
		6	1 or 2	1633	2633
4	36	-			
7	29	-			
9	01	6	1 or 2	1699	2699

Of the options for *KISS*, only 2033 squares to the right kind of *PASSION*, 413,3089.

The Twelve Days of Christmas (p. 148)

1. Five gold rings each on eight days = 40
Eight maids for each of five days = 40

2. Six geese a-laying are given on seven days, and seven swans a-swimming are given on six days; hence a total of 42 geese and 42 swans are given. The largest total of birds, over the 12 days of Christmas, is 42 geese a-laying, equaled by 42 swans a-swimming.

3. 184 (= 1 × 12 + 2 × 11 + 3 × 10 + 4 × 9 + 6 × 7 + 7 × 6).
1 partridge on 12 of the days
2 turtledoves on 11 of the days
3 French hens on 10 of the days
4 calling birds on 9 of the days
6 geese on 7 of the days
7 swans on 6 of the days

4. Five gold rings are given for the first time on the fifth day. And five gold rings were given on eight days. In general, if *n* items are given for the first time on the *n*th day, then those items are given on (13 − *n*) days. Hence,

$$n(13 - n) = 13n - n^2$$

of a particular item will be given when the item is given for the first time on day *n*.

5. 78; 364.
On the 13th row, go over three places, and read 78.
Go to the 14th row and find 364.
Note that to compute $\binom{n}{r-1}$, we get the *r*th entry in the *n*th row in the Pascal's triangle.

6. Answers vary.

7. 54 diagonals.

Number of sides of polygons	Number of diagonals
3	0
4	2
5	5
6	9
⋮	⋮
12	54

8.

Number of sides of polygons (x)	Number of diagonals ($ax^2 + bx + c$)
3	$9a + 3b + c$
4	$16a + 4b + c$
5	$25a + 5b + c$
6	$36a + 6b + c$
7	$49a + 7b + c$
⋮	⋮

Applying the method of finite differences to both the general case and the specific case, and comparing both sides, we have

$$2a = 1 \Rightarrow a = \frac{1}{2}$$

$$7a + b = 2 \Rightarrow 7\left(\frac{1}{2}\right) + b + 2$$

$$\Rightarrow b = -\frac{3}{2}$$

$$9a + 3b + c = 0$$

$$9\left(\frac{1}{2}\right) + 3\left(-\frac{3}{2}\right) + c = 0$$

$$c = 0$$

Thus the quadratic that generates the number of diagonals in any polygon is

$$\frac{x^2}{2} - \frac{3x}{2} = \frac{x(x-3)}{2}$$

where x is the number of sides.

9. Observe that $\binom{n+2}{3} = \frac{(n+2)(n+1)n}{1 \times 2 \times 3}$

$$= \frac{n^2}{6} + \frac{n^2}{2} + \frac{n}{3}.$$

Diagonals in a Polygon

General			Specific	
$9a + 3b + c$			0	
	$7a + b$		2	
$16a + 4b + c$		$2a$	2	1
	$9a + b$		3	
$25a + 5b + c$		$2a$	5	1
	$11a + b$		4	
$36a + 6b + c$		$2a$	9	1
	$13a + b$		5	
$49a + 7b + c$			14	
⋮		⋮		

Bibliography & References

Acertijeros, L. (1992). Quadrennially. *Journal of Recreational Mathematics*, *24*(1), 52.

Adler, A. & Coury, J. E. (1995). *The theory of numbers: A text and source book of problems*. Sudbury, MA: Jones and Bartlett Publishers.

Andel, J. (2001). *Mathematics of chance*. New York: John Wiley & Sons, Inc.

Ash, R. (2007). *Potty, Fartwell & Knob: Extraordinary but true names of British people*. London: Headline Publishing Group.

Barwell, B. (1992). DAY + DAY +···+ DAY = YEAR. *Journal of Recreational Mathematics*, *24*(1), 54.

Beiler, Albert H. (1966). *Recreations in the theory of numbers: The queen of mathematics entertains*. New York: Dover Publications, Inc.

Berloquin, Pierre (1980). *Games of logic*. London: Unwin Paperbacks.

Berners-Lee, M. (2010). *How bad are bananas?* London: Profile Books Ltd.

Bolton, M. & Dickson, B. (2007). *Maybe life's just not that into you*. New York: Howard Books.

Book, D. L. (1992). *Problems for puzzlebusters*. Washington, DC: Enigmatics Press.

Botham, N. (2009). *The pocket book of useless information*. London: John Blake Publishing Ltd.

Bradbury, A. G. (2003–2004). Alphametics: 2597 Garden party – 1. *Journal of Recreational Mathematics*, *32*(3), 245.

Brasch, R. (1996). *Mistakes, misnomers and misconceptions*. Sydney: Angus&Robertson.

Buckeridge, J., Gulyaev, S. & Klymchuk, S. (2005). *Shape puzzles*. Singapore: Times Editions.

Bunch, B. (1982). *Mathematical fallacies and paradoxes*. New York: Dover Publications, Inc.

Burroughs, A. (2009). *You better not cry: True stories for Christmas*. London: Atlantic Books.

Burton, R. A. (2008). *On being certain*. New York: St. Martin's Press.

Covill, R. J. (1993). Old testament. *Journal of Recreational Mathematics*, *25*(4), 294.

Chua, S. K. et al. (2007). *Singapore mathematical olympiads (1995-2004)*. Singapore: Singapore Mathematical Society.

Derrick, N. (2009). *Number freak*. New York: A Pedigree Book.

Devlin, Keith (2000). The mathematics of christmas. *Devlin's Angle*. (December 2000). http://www.maa.org/devlin/devlin_12_00.html

Dudeney, H. E. (1958). *Amusements in mathematics*. New York: Dover Publications, Inc.

Dugle, J. (1982). The Twelve Days of Christmas and Pascal's Triangle. *Mathematics Teacher*, *75*(9), 755-57.

Elffers, J. (1973). *Tangram: The ancient Chinese shapes game*. Penguin Books.

Eves, H. E. (1988). *Return to mathematical circles*. Boston: PWS-KENT Publishing Company.

Fixx, J. F. (1972). *Games for the superintelligent*. New York: Galahad Books.

Frederickson, G. N. (1997). *Dissections: Plane and fancy*. Cambridge: Cambridge University Press.

Frohlichstein, J. (1962). *Mathematical fun, games and puzzles*. New York: Dover Publications, Inc.

Gardiner, A. (1987). *Mathematical puzzling*. New York: Dover Publications, Inc.

Gardner, M. (2006). *The colossal book of short puzzles & problems*. New York: W. W. Norton & Company, Inc.

Hartley, W. J. (1986). *Christmaths: Games, investigations, puzzles*. Leicester: The Mathematical Association.

Henry, J. B. et al. (1997). *Challenge! 1991-1995*. Australian Mathematics Trust.

Highfield, R. (1998). *Can reindeer fly?: The science of christmas*. London: Metro Books.

Ho, J. B. (1991). *Secondary mathematics 1A: Teacher's guide*. Singapore: Curriculum Development Institute of Singapore/Pan Pacific Publications Ltd.

Hopkins, N. J., Mayne, J. W. & Hudson, J. R. (1992). *The numbers you need*. London: Gale Research Inc.

Hunter, J. A. H. & Madachy, Joseph S. (1963). *Mathematical diversions*. New York: Dover Publications, Inc.

Jargocki, C. P. (1976). *Science brain-twisters, paradoxes, and fallacies*. New York: Charles Scribner's Sons.

Kador, J. (2004). *How to ace the brain teaser interview*. New York City: McGraw-Hill.

King, L. (2003). *The Times test your creative thinking*. London: Kogan Page Limited.

Knowles, D. H. (1979). Fun at Christmastime. *Mathematics Teacher*, *72*(9), 669-673.

Kordemsky, B. A. (1972). *The Moscow puzzles*. Penguin Books.

Lavin, M. (ed.). (2004). *The business of holidays*. New York: The Monacelli Press.

McMaster, A. (1986). The Twelve Days of Christmas and the Number of Diagonals in a Polygon. *Mathematics Teacher 79*(9), 700-702.

Morgan, F. (2000). *The math chat book*. The Mathematical Association of America.

Moscovich, I. (2001). *1000 playthinks: Puzzles, paradoxes, illusions & games*. New York: Workman Publishing Company, Inc.

Mott-Smith, Geoffrey (1954). *Mathematical puzzles for beginners and enthusiasts*. New York: Dover Publications, Inc.

Nelson. H. L. (1982–1983). Yuletide sentiment. *Journal of Recreational Mathematics, 14*(3), 220.

Niederman, D. (2009). *Number freak*. New York: A Perigee Book.

Peekham, A. (2007). *Mo' Urban dictionary*. Kansas City: Andrews McMeal Publishing, LLC.

Peekham, A. (2005). *Urban Dictionary: Fularious street slang defined.* Kansas City: Andrews McMeal Publishing, LLC.

Phillips, R. (1994). *Numbers: Facts, figures and fiction*. Cambridge University Press.

Pöppelmann, C. (2006). *1,000 common delusions and the real facts behind them*. New York: Firefly Books.

Poundstone, W. (2003). *How would you move Mount Fuji?* New York: Little, Brown.

Radtke, J. R. (1979?). Holiday graphs. *Mathematics Teacher, 74*(9), 680.

Rae, S. (Ed.) (1996). *The Faber book of Christmas*. London: Faber and Faber Limited.

Raphel, A. (1993). Critical thinking during the December holidays. *Arithmetic Teacher, 41*(4), 216-219.

Read, R. C. (1965). *Tangrams – 300 puzzles*. New York Dover Publications, Inc.

Schimmel, A. (1993). *The mystery of numbers*. New York: Oxford University Press.

Sharpe, G. & Schlaifer, R. L. (2007). *What are the odds?* London: Orion Books.

Shirali, S. (2002). *A primer on logarithms*. India: Universities Press.

Standiford, L. (2008). *The man who invented Christmas: How Charles Dickens's* A Christmas Carol *rescued his career and revived our holiday spirits*. New York: Crown Publishers.

Steinwachs, R. (2009). *Super brain bafflers*. New York: Sterling Publishing Co., Inc.

Stueben, M. (1984). Brain Bogglers. *Discover*, December 1984, pp. 100, 98.

Stewart, I. (1997). *The magical maze: Seeing the world through mathematical eyes*. New York: John Wiley & Sons, Inc.

Tattersall, J. J. (2005). *Elementary number theory in nine chapters (2nd ed.)*. Cambridge: Cambridge University Press.

Thompson, S. (2009). *Trading secrets: 20 hard and fast rules to help you beat the stock market*. Harlow, England: FT Prentice Hall.

Trigg, C. W. (1982/83). Training zeros in factorials. *Journal of Recreational Mathematics, 15*(1), 57.

Van Delft, P. & Botermans, J. (1995). *Creative puzzles of the world*. Berkeley, California: Key Curriculum Press.

Van Note, P. (1968). *The 8th book of Tan: 700 tangrams by Sam Loyd*. New York: Dover Publications, Inc.

Vance, I. E. (1981). A partridge in a pear tree, a stack of cubes, and four buckets of balls. *Mathematics Teacher, 74*(9), 698-703.

Vennebush, G. P. (ed.) (2004). The 12 days of Christmas costly. *Mathematics Teacher, 98*(5), 322-324.

Waldfogel, J. (2009). *Scroogenomics: Why you shouldn't buy presents for the holidays.* Princeton, New Jersey: Princeton University Press.

Weinstein, L. & Adam, J. A. (2008). *Guesstimation: Solving the world's problems on the back of a cocktail napkin.* New Jersey: Princeton University Press.

Wells, D. (1991). *The Penguin dictionary of curious and interesting geometry.* London: Penguin Group.

Wells, D. (1986). *The Penguin dictionary of curious and interesting numbers.* London: Penguin Books.

Welchman-Tischler, Rosamond (1992). *The mathematical toolbox.* New York: Cuisenaire Company of America, Inc.

Wertheim, M. (1995). *Pythagoras' trousers: God, physics, and the gender wars.* New York: Times Books.

Wright, S. (2007). *Steve Wright's further factoids.* London: Harper Collins Publishers.

Yan, K. C. (2010). *More mathematical quickies & trickies.* Singapore: MathPlus Publishing.

Yan, K. C. (2010). *Mathematical quickies & trickies (expanded & updated).* Singapore: MathPlus Publishing.

Yan, K. C. (2008). *Mind stretchers 2.* Singapore: Panpac Education.

Yan, K. C. (2006). *Aha! Math.* Singapore: EPB Panpac.

Yan, K. C. (2004). Joyeux ChristMATHS. *YG Singapore*, 284, 16-17.

Yan. K. C. (2003). A formula for Christmas day. *YG Singapore*, 272, 12.

Yan, K. C. (2003). The mathematics of Christmas. *YG Singapore*, 272, 10-11.

Yan, K. C. (2003). The 12 puzzles of Christmas. *YG Singapore*, 272, pp. 6-7.

http://mathworld.wolfram.com/OctahedralNumber

http://math.world.wolfram.com/CullernNumber

http://mathworld.wolfram.com/HilbertNumber

www.urbandictionary.com